The Mystery of Carbon

An introduction to carbon materials

The Mystery of Carbon

An introduction to carbon materials

Manijeh Razeghi
Center for Quantum Devices, Northwestern University, Evanston, Illinois, USA

IOP Publishing, Bristol, UK

ISBN 978-0-7503-1182-3 (ebook)
ISBN 978-0-7503-1183-0 (print)
ISBN 978-0-7503-1184-7 (mobi)

DOI 10.1088/2053-2563/ab35d1

Version: 20191101

IOP ebooks
ISSN 2053-2563 (online)
ISSN 2054-7315 (print)

British Library Cataloguing-in-Publication Data: A catalogue record for this book is available from the British Library.

Published by IOP Publishing, wholly owned by The Institute of Physics, London

IOP Publishing, Temple Circus, Temple Way, Bristol, BS1 6HG, UK

US Office: IOP Publishing, Inc., 190 North Independence Mall West, Suite 601, Philadelphia, PA 19106, USA

To my family, my past, present and future students.

Contents

Preface

This book is called *The Mystery of Carbon* and its aim is to give solid-state engineering students a brief and concise introduction to the very exciting and powerful recent discoveries made in the field of carbon materials, their synthesis, allotropes, and the impact this has had on developmental science. There are in the meantime many new books dedicated to carbon and carbon technology. The reader can review the recent publications by using Google and searching for 'carbon' and related materials such as 'carbon nanotubes', 'fullerenes', 'graphene', 'graphite', etc. This monograph differs from the above mentioned books and reviews in one important aspect: it is designed for the benefit of the solid-state and electrical engineering student at graduate level. The objective is to give the student a quick introduction to the recent discoveries made with carbon and show him/her how these new effects and materials could be useful to his/her own work and creative thinking, without having to pass through the vast amount of organic and physical chemistry which is normally bundled together into a book on carbon. At the same time, the book is not written like a conference proceeding or review collection written by specialized authors. The latter get to the point quickly, but they are often too specialized, and too difficult for a graduate student to follow, unless she/he is already working in that field of research. The material covered by this book is far from complete; the applications discussed for the various materials are very limited to the applications which are of direct relevance to solid-state engineering programs, with emphasis on the new science and technology of nanostructures and nanodevices.

Chapter 1 is an introduction to carbon, with a focus on carbon as an element, its atomic structure, and the ability that the carbon atom has to bond in a variety of configurations, which is what gives the element the key role that it has in chemistry, biology and life. Chapter 2 is dedicated to the properties of pure tetrahedral carbon, or diamond, and some of its applications to engineering, which includes the role of N-defects and doping, leading to quantum information applications and superconductivity, respectively. Chapter 3 is devoted to buckminsterfullerenes, or Buckyballs. This is a material which led to a Nobel Prize for Harry Kroto, who found that with potassium doping, K_3C_{60} is a superconductor with the fabulous transition temperature of $\sim$40 K, unprecedented for an organic material, and even higher than almost all inorganic materials. This discovery stimulated researchers and engineers to take organic semiconductor physics very seriously, and to pay attention to its potential, far more than had been done so far. Suddenly it was thought that maybe it is the organics that hold the key to new technologies, and considerable efforts and resources were freed to follow this up. That actually turned out to be true. However, it was not high-T_c superconductivity that was discovered, but rather, the ultra-high mobility, and amazing topologies, and order, which were uncovered with the synthesis and extraction of monolayer graphene (abbreviated G or MLG), and then later bilayer graphene (BLG). MLG and BLG and some applications thereof are discussed in chapter 4. The description and details provided in chapter 4 are very brief and one cannot do justice to this incredibly important area of science and technology in a short review. The material in chapter 4 shows the engineering student why G implies a revolution in technology, and why she/he should look at its properties and applications in nano-engineering device design. The discovery of graphene gave Novoselov and Geim the Nobel Prize in 2010. In chapter 5 we discuss the properties and applications of carbon nanotubes (abbreviated as CNTs). CNTs can be viewed as 'rolled up' forms of graphene. Rolling up the G-hexagonal carpet can be done in various angles and this leads to different chiral symmetries with very different band structures. The art of CNT synthesis has been developed to reach very practical schemes, and scientists now know how to tune bandgaps and make single- or multi-walled CNTs on demand. CNTs have some of the best mechanical and structural stabilities and can be used in large-scale engineering applications such as building materials, with suitable thermal and electrical conducting properties, and in photovoltaic cells. The field of conjugated polymers is described in chapter 6. This is a fascinating subject which occupied the minds of many scientists in the 1980s. Alan Heeger and his Santa Barbara Group discovered that trans-polyacetylene was a Peierls semiconductor which spontaneously undergoes a structural phase transition triggered by an electron–phonon interaction along the backbone. The transition also gives rise to beautiful new science, and a new kind of collective excitations known as polarons and solitons. These were first predicted and then measured. The discoveries gave Heeger the 2000 Nobel Prize. Then in chapter 7 we present some very new and brilliant discoveries, that, even though not strictly related to carbon, exhibit almost monolayer 2-dimensionality (2D), which is close in structure to graphene. These 2D transition metal dichalcogenides (TMD), can now, with present know-how, be grown as multilayers, mixed with graphene layers and insulating barriers –BN layers. The science of van der Waals and 2D-TMD layer-by-layer superlattice physics is now born. The discoveries in this field are just beginning, and one can sense that physicists are hoping this could be the magical material systems which can eventually achieve the ultimate goal: making high-T_c superconductors with critical temperatures above 120 K and efficient solar cells. But instead, these materials are solving another great challenge in technology: making sensitive and commercial gas, molecular and bio-sensors.

The book material is again summarized in chapter 8 with conclusions and perspectives.

My hopes and expectations are that the graduate student can read this book, maybe listen to it in a class as part of his/her basic training, and that he/she will enjoy the new science and understand it without a great deal of effort. The material is not very demanding and should provide a first quick introduction to the subject.

Finally, I would like to acknowledge the contribution of the following: Samy Annabi, Victor XU, Ferechteh Hosseini Teherani, Bijan Movaghar, Pedram Khalili, Ryan McClintock and Fehu Wang.

Manijeh Razeghi
Walter P. Murphy Professor
Director and Founder of the Center for Quantum Devices,
Department of Electrical and Computer Engineering
Northwestern University Evanston 2019

Prerequisite

The reader should have a knowledge of solid-state physics at the undergraduate level (from books such as those by C Kittel, Ashscroft and Mermin, J Ziman, O Madelung). Preferably she/he should be acquainted or have at hand the book by Manijeh Razeghi entitled *Fundamentals of Solid State Engineering (*FSSE), second, third or fourth edition, published by Springer. She/he should be aware of **second quantization and density functional theory (DFT)**. It was deemed best to give good references to articles covering these last two mentioned topics rather than deriving them in an appendix. The articles cited are available online.

Foreword

Why is this book on carbon so important as a must-read text for all? Carbon, derived from the Latin root, coal, is and will continue to be a large component of our energy source for the next several hundred years as projected by the US Department of Energy. It is also noted that carbon forms millions of compounds with other elements that are crucial to human lives. Important examples include carbon dioxide, which is used in photosynthesis when combined with water and sunlight.

Professor Manijeh Razeghi has done a challenging task in this book by integrating the fundamental science of carbon with current research topics to stimulate the interest of her readers at the senior/graduate level. This short and concise book takes only one sitting to read, but one walks away with a global perspective on the subject. I strongly recommend all to acquire from this book some understanding of how carbon is impacting our lives.

I sincerely thank Manijeh for her dedication and contribution to our knowledge-base.

R P H Chang
Professor of Materials Science and Engineering
Northwestern University

Author biography

Manijeh Razeghi

Manijeh Razeghi received the Doctorate d'état ES Sciences Physiques from the Université de Paris, France, in 1980. After heading the Exploratory Materials Lab at Thomson-CSF (France), she joined Northwestern University, Evanston, IL, as a Walter P. Murphy Professor and Director of the Center for Quantum Devices in Fall 1991, where she created the undergraduate and graduate program in solid-state engineering. She is one of the leading scientists in the field of semiconductor science and technology of quantum materials and devices. She has authored or co-authored more than 1000 papers, more than 32 book chapters, and 19 books, and has given more than 1000 invited and plenary talks. Her current research interest is in nanoscale optoelectronic quantum devices and systems from deep UV (100 nm) to Thz (300 microns).

Dr Razeghi is a Fellow of MRS, IOP, IEEE, APS, SPIE, OSA, Fellow and Life Member of Society of Women Engineers (SWE), Fellow of the International Engineering Consortium (IEC), and a member of the Electrochemical Society, ACS, AAAS. She received the IBM Europe Science and Technology Prize in 1987, the Achievement Award from the SWE in 1995, the R.F. Bunshah Award in 2004, IBM Faculty Award 2013, and the Jan Czochralski Gold Medal in 2016, and the Benjamin Franklin Award in Electrical Engineering in 2018, as well as many best paper awards.

She is an elected Life Fellow of SWE, IEEE, and MRS.

IOP Publishing

The Mystery of Carbon
An introduction to carbon materials
Manijeh Razeghi

Chapter 1

The carbon atom

1.1 Introduction

This overview aims to examine the basic characteristics of the carbon element and to review this atom's fundamental properties, especially its electronic structure and energy levels. We will then understand the amazing versatility of the carbon bond enabling the formation of various chemical complexes and allotropes. The most important class of carbon allotropes is enumerated and the structures shown. Carbon isotopes are also examined. The remaining material in this chapter is dedicated to the question of carbon bonding and energy bands in solids. The theory of energy bands in solids, including the density functional (DFT) and tight binding methods (TBM), is developed in many excellent articles, some of which are cited in this review.

The carbon atom

Carbon is without a doubt one of the most common elements encountered in the universe: following hydrogen, helium and oxygen, it is indeed the fourth most abundant element by mass. Carbon is involved in every known life form; it is notably the second most abundant element by mass in the human body after oxygen (about 18.5%). This abundance, combined with the extraordinary variety of carbon-based organic compounds and with its unusual ability to form polymers at naturally encountered Earth temperatures, make this element the chemical foundation of all life as it is known [1–3]. The carbon atom forms a very large number of components. It is astonishing to note that there are, to our knowledge, so far one million organic compounds containing only carbon and hydrogen.

In the periodic table (see figure 1.1), carbon is in group 14, and is therefore tetravalent (four electrons are disposable to compose covalent chemical bonds) and not a metal. This tetravalence explains the formation of carbon compounds and is the origin of carbon's abundance in life. This atom has indeed the outstanding feature to form single bonds with itself that can resist most reactions occurring at

doi:10.1088/2053-2563/ab35d1ch1

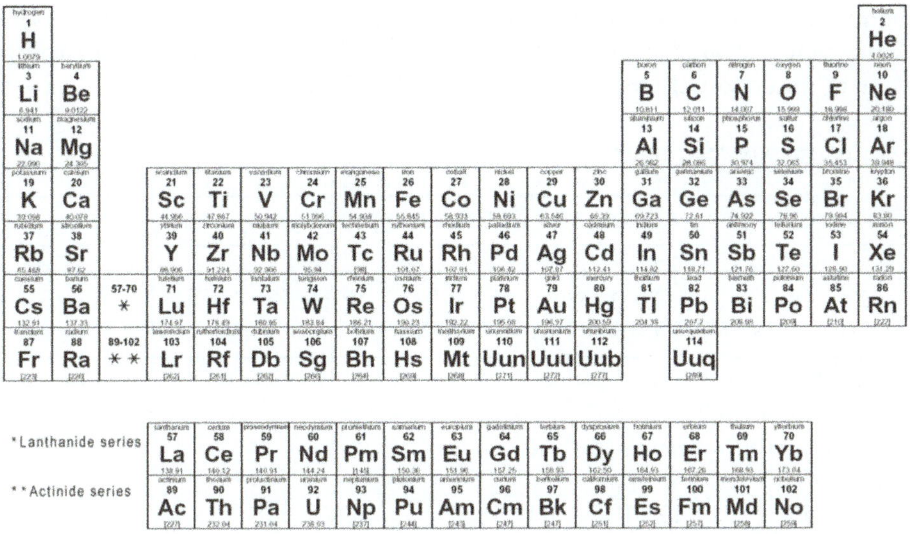

Figure 1.1. The periodic table, carbon is located in group 14. Reprinted from [4] by permission from Springer Nature, copyright 2019.

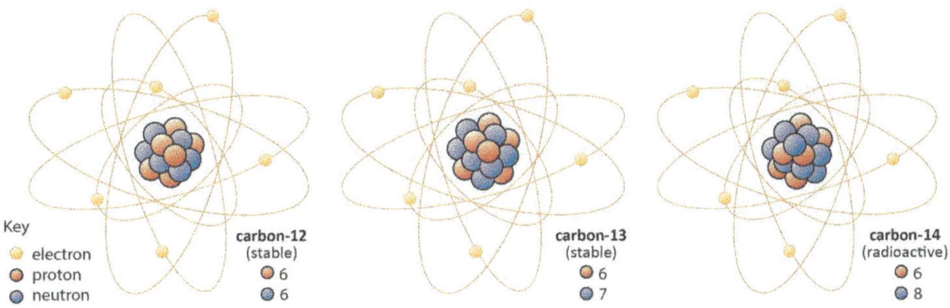

Figure 1.2. The three natural isotopes of carbon. Reprinted from [4] by permission from Springer Nature, copyright 2019.

ambient conditions. Long chains of carbon atoms can therefore be formed, constituting the basis of living cell's compounds, such as DNA (figure 1.2).

Isotopes of carbon

Carbon has 15 known isotopes. Their atomic mass ranges from 8 to 22 (^{8}C to ^{22}C), where only the number of neutrons varies. The only stable isotopes are ^{12}C and ^{13}C, the others are radioactive. The most stable radioisotope is ^{14}C: its half-life is about 5730 years, while it does not exceed 20 s for the other ones. In this textbook, unless mentioned otherwise, the properties described and calculations are assumed to be done for the ^{12}C isotope, as it is the most abundant isotope in nature. Thanks to mass spectroscopy measurements, we now know that ^{12}C isotope constitutes 99% of the whole carbon population. It is interesting to see that ^{12}C has been chosen to define the unified atomic mass unit, the Dalton: one Dalton is equal to one twelfth of the mass of an unbound atom of ^{12}C ($1.660\ 539\ 066\ 60 \times 10^{-27}$ kg).

Isotopes of carbon characteristics

	Z	N	Mass (u)	Half-life (y)	Decay mode	Nuclear spin
^{12}C	6	6	12	Stable		0+
^{13}C	6	7	13.003 355	Stable		½−
^{14}C	6	8	14.003 242	5730	β−	0+

[References 5–7]

Some fundamental properties of ^{12}C
Atomic mass: 6
Atomic weight: ^{12}C is: 12.0107 ± 0.0008 u (1u = $1.660\ 538\ 92 \times 10^{-27}$ kg)
Van der Walls radius: 170 pm
(From [5–7])

Electronic configuration
The electronic configuration of carbon is [$1s^2$] $2s^2$ $2p^2$. The n = 1 shell is thus fully filled with two electrons, the n = 2 shell is filled by two electrons in the s-subshell (full) and two electrons in the p-subshell (not full as there is room for six electrons). A reminder of how electronic configuration is obtained can be found in *Fundamentals of Solid State Engineering* (FSSE) by M Razeghi ('Atomic Orbital', chapter 1 of [4]).

The atom's reactivity is set by the surface shell, so hereby the n = 2 shell. The binding capacities of the carbon atom can be explained by studying this specific energy level. To enable carbon atoms to bond with each other and to other compounds, one $2s^2$ electron can be promoted to the p-subshell: the electronic configuration therefore becomes 2s $2p^3$. However, this transfer costs energy as the p-subshell energy level is higher than the s-subshell one (figure 1.3), so the formed bond must counter this loss.

Figure 1.4 shows the formation of the three-dimensional diamond lattice structure where four covalent bonds are formed in a tetrahedral configuration by combining a 2s electron with three 2p ones. This so-called sp^3 arrangement optimizes the overlap of opposite charges, also shrinking the energy far more than the cost of the $2s^2$ electron promotion to the p-subshell.

Carbon also has the possibility to bond in a two-dimensional configuration, such as in graphene. In this arrangement three sp^3 bonds are formed leaving one p-orbital standing. Finally, the sp^3 can also link in a linear configuration, forming only two linear bonds and leaving two p-orbitals standing. This will be explained in detail later.

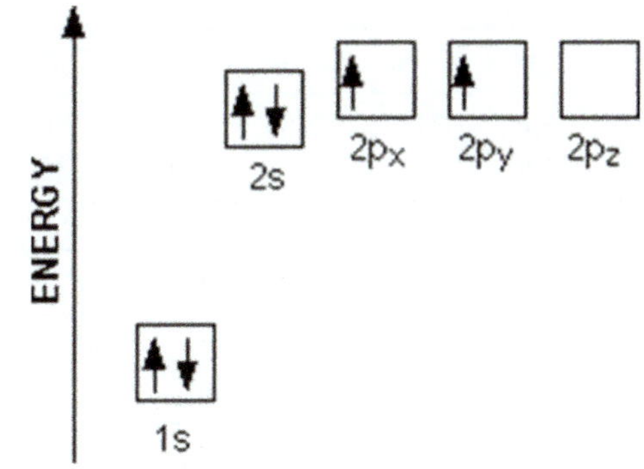

Figure 1.3. The energy level structure of the carbon atom. Hund's rule gives a triplet ground state: when there is space in the shell electron/electron repulsion favors the triplet state. Reprinted from [4] by permission from Springer Nature, copyright 2019.

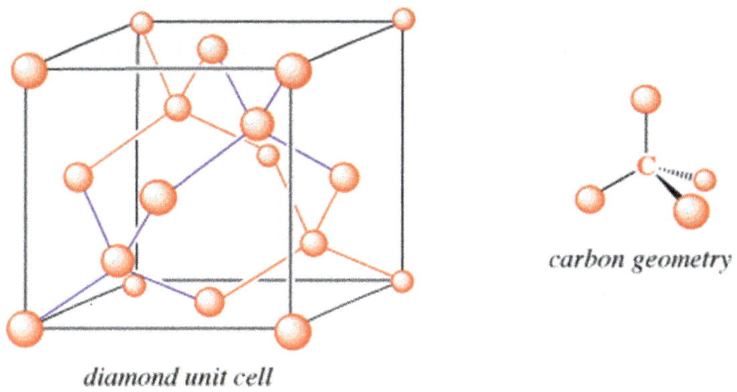

Figure 1.4. The carbon tetrahedral sp^3 bonds and the formation of the covalent three-dimensional sp^3 covalent diamond lattice. Reprinted from [4] by permission from Springer Nature, copyright 2019.

The bond length of a tetrahedral configuration, also called the covalent radius, is about 77 pm, while it shrinks to 73 pm in graphene and to 69 pm in a linear arrangement. The first ionization energy of carbon (the energy required to excite an electron and remove it from the atom in the gas phase) is 1086.5 kJ mol^{-1}.

Binding energies

This table provides useful constants concerning covalent bindings involving the carbon atoms: the bonding energy D, which corresponds to the energy required to break the bond, and the bonding length R, which is the distance between two atoms engaged in a bonding.

Bond	C–C	C=C	C≡C	C–Si	C–Ge	C–Sn	C–Pb	C–N	C=N	C≡N
D (kJ mol^{-1})	346	602	835	318	238	192	130	305	615	887
R (pm)	154	134	120	185	195	206	230	147	129	116

C–P	C–O	C=O	C≡O	C–B	C–S	C=S	C–F	C–Cl	C–Br	C–I
264	358	799	1072	356	272	573	485	327	285	213
184	143	120	113		182	160	135	177	194	214

[8, 9]

1.2 Electronic configuration and covalent bonding between carbon atoms

Similar to other elements such as Si ($3s^23p^2$) or Ge ($4s^24p^2$), an electron from the filled s-orbital can be promoted to the p-subshell of the carbon ground state ($2s^22p^2$) to enable covalent bonds that lower the energy and balance the cost of the initial uplift 's to p'. Carbon, however, distinguishes itself by not only being unrestrained from forming tetrahedral bonds, but also planar ones with three sp^2 covalent bonds and a perpendicular p-orbital, and linear ones (sp bond). These different configurations have many consequences, which we will discuss in this chapter.

The most common form of the carbon–carbon bond (the covalent bond between two carbon atoms) is the single bond, also called a sigma bond. It is composed of one electron from each atom and is thus a two-electron bond. Each atom contributes with one hybridized orbital, for example the sp^3 ones in ethane C2H6. Other kinds of hybridizations can also be used to create single bonds, for instance sp^3 to sp^2: there is no hybridization requirement to bond two carbon atoms (figure 1.5). Double bonds also exist between two carbon atoms, generating the alkene group, also called olefins (see figure 1.6). These hydrocarbons are composed exclusively of carbon and hydrogen, are unsaturated and present at least one C–C double bond. Two carbon atoms can also be triple-bonded in compounds called alkynes.

To form a double bond, both an sp^2 hybridized orbital and an unhybridized p-orbital are involved. For triple bonds, an sp-hybridized orbital and two p-orbitals from each atom are engaged. Bonds formed using p-orbitals are called pi bonds.

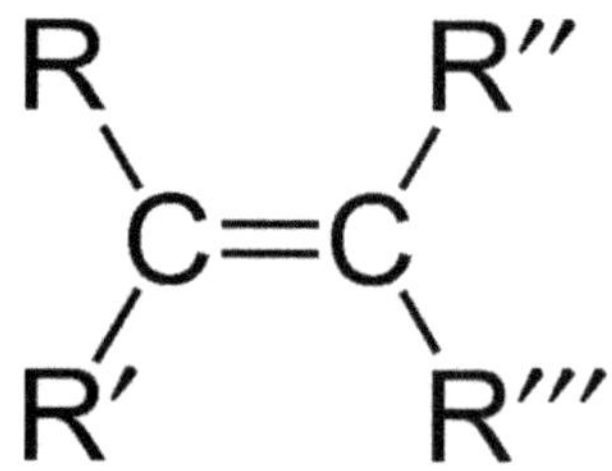

Figure 1.5. Alkene flat. The R R′ are any side groups that attach to C with a single bond and fit into space. Reprinted from [4] by permission from Springer Nature, copyright 2019.

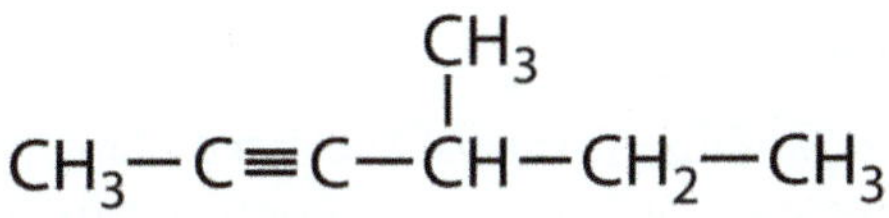

Figure 1.6. 4-Methyl-2-hexyne molecule, example of a triple-carbon bond. Reprinted from [4] by permission from Springer Nature, copyright 2019.

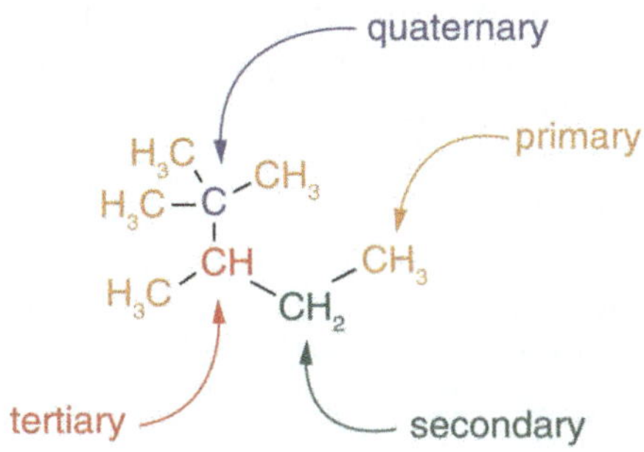

Figure 1.7. Trimethylpentane, example of the versatile way carbon can bond to itself and other compounds. Reprinted from [4] by permission from Springer Nature, copyright 2019.

Carbon is an element that can form long chains with its own atoms, a feature called catenation. It is one of the few elements that can do that. Combining this capacity with the strength of the carbon–carbon bond makes it possible to generate a very large number of molecular compounds. Of these compounds many are important constituents of life. This is the reason why there is a specific field of study on carbon compounds called organic chemistry.

In C–C skeleton structures branching is common (figure 1.7). One can identify different carbon atoms according to the number of its carbon neighbors:

- **primary carbon atom**: one carbon neighbor
- **secondary carbon atom:** two carbon neighbors
- **tertiary carbon atom:** three carbon neighbors
- **quaternary carbon atom:** four carbon neighbors

Carbon shows a substantial versatility and flexibility that allows it to bond in various arrangements, thus making this atom a singular and unique element, and a crucial constituent of life.

The following aims at reviewing the structures of major carbon allotropes, which is the main objective of this chapter. This review cannot pretend to cover the broad field of organic chemistry and of carbon physics, but some references will be provided when particularly interesting scientific issues arise. The next chapters will be devoted to pure carbon allotropes and to introducing their basic properties (electronic, optical, etc), as well as their singularity and applications.

With what has already been introduced, we can already provide a possible explanation to the mystery of carbon: it should be related to the sp, sp^2 and sp^3 carbon–carbon bond.

1.3 Carbon allotropes

Allotropes are different crystalline forms of the same element in the periodic table.

Among all carbon allotropes, the three most prominent ones are graphite, diamond and lonsdaleite. Each allotropic form has its own physical properties, which can be very different from one allotrope to the other. Current materials research is indeed focusing on examining the characteristics of these various crystalline forms. A striking example is the comparison between diamond and graphite. The first one is highly transparent, while the latter presents a dark opacity and a black aspect. Diamond is the hardest naturally occurring material, while graphite is soft and leaves a trail on paper. The electrical conductivity of diamond is very low, whereas graphite conducts perfectly well. Some carbon-based allotropes such as diamond, carbon nanotubes and graphene present the highest thermal conductivities of all materials under standard conditions. Power electronics require high-thermal conductivity materials, for instance to build high-performing computers, such as the trillion-dollar personal computer and laptop industry demands (figure 1.8).

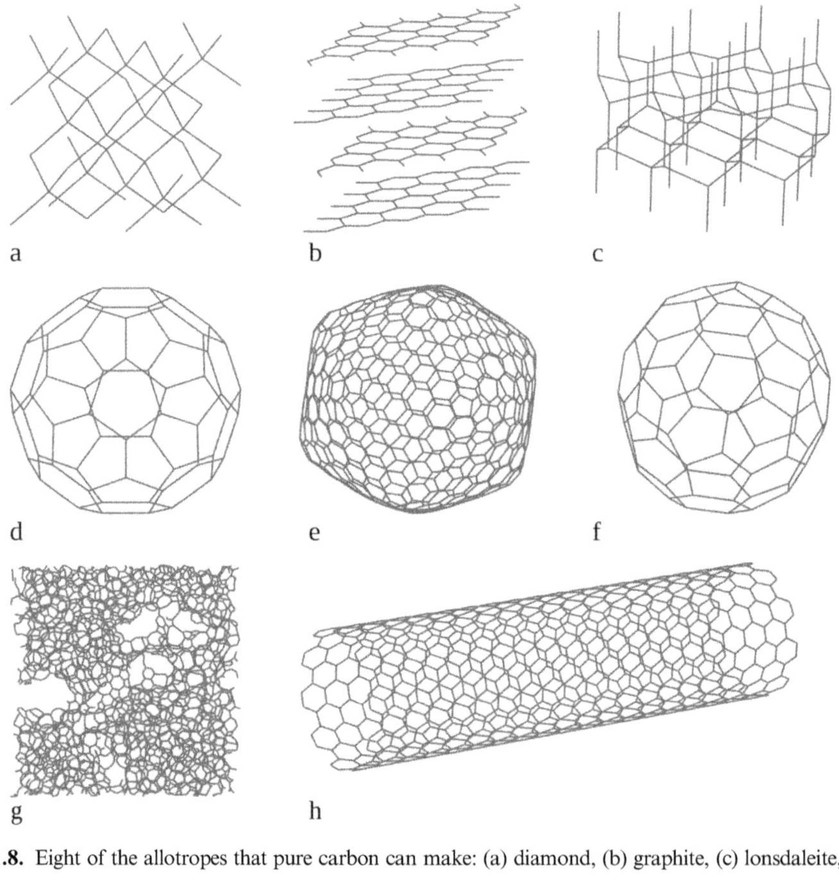

Figure 1.8. Eight of the allotropes that pure carbon can make: (a) diamond, (b) graphite, (c) lonsdaleite, (d) C60 (buckminsterfullerene), (e) C540 (see fullerene), (f) C70 (see fullerene), (g) amorphous carbon, (h) single-walled carbon nanotube. This image has been obtained by the authors from the Wikimedia website where it was made available by Andel under a CC BY-SA 4.0 licence. It is included within this book on that basis. It is attributed to Andel.

Some allotropes of carbon:

(a) Diamond, tetrahedral bonding sp^3 (discussed in chapter 2).
(b) Graphite, two-dimensional sp bonding combined with a van der Waals bonded layered structure (weak bonding via mutually induced dipole–dipole coupling). Graphene can be obtained by extracting a two-dimensional layer, as discussed in chapter 4.
(c) Lonsdaleite, an allotrope with a hexagonal lattice, also called hexagonal diamond, and named after Kathleen Lonsdale. It can be found in nature and is known to form when meteorites containing graphite strike the Earth. It is a translucent material with a brownish-yellow aspect, an index of refraction between 2.40 to 2.41 and a specific gravity of 3.2 to 3.3.
(d) Fullerenes, carbon-based sphere (discussed in chapter 4).
(e) Amorphous carbon, where hydrogen can saturate dangling bonds for instance.
(f) Carbon nanotube, carbon-based tube (discussed in chapter 5).

This table proposes a comparison between some carbon allotropes, using their electronic configuration and structural properties:

Allotrope	Hybridization	Structure	Existence
Graphite	sp^2	Crystal, 2D hexagonal stacked	Natural
Graphene	sp^2	Crystal, hexagonal stacked	Natural or synthetic (monolayer)
Diamond	sp^3	Crystal, Cubic	Natural
Lonsdaleite	sp^3	Crystal, 3D hexagonal	Natural
Fullerene	sp^2	Cluster	Synthetic
Nanotube	sp^2	With single, double or multiple walls	Synthetic
Amorphous carbon	sp^2–sp^3	No crystalline structure	Natural

A comparison of graphite and diamond

	Graphite	Diamond
Mechanical hardness	Graphite is one of the softest materials.	Synthetic nanocrystalline diamond is the hardest material known.
Lubricant properties	Graphite is a very good lubricant.	Diamond is the ultimate abrasive.
Electrical conduction	Graphite is a conductor of electricity.	Diamond is an excellent electrical insulator and has the highest breakdown electric field of any known material.
Thermal conduction	Graphites are used for thermal insulation but some other forms are good thermal conductors.	Diamond is the best known naturally occurring thermal conductor.
Optical transparency	Graphite is opaque.	Diamond is highly transparent.
Lattice structure	Graphite crystallizes in the hexagonal system.	Diamond crystallizes in the cubic system.

[9]

This simple comparison between graphite and diamond reveals how structural differences can lead to drastically different basic properties, even with the same atom crystallizing in different arrangements. These properties remain until now unexplained, but there is a certain enthusiasm to discover new crystalline structures to find astounding results—as with graphene, which is now a subject of intensive research. The following aims to examine in detail these structures and properties.

This table gives some numerical values of these properties, revealing the great variety of properties that can be expected from carbon allotropes:

	Density ($g\ cm^{-3}$)	Molar heat capacity ($J.mol^{-1}\ K^{-1}$)	Thermal conductivity ($W\ .\ m^{-1}\ K^{-1}$)	Mohs hardness	Electrical resistivity (Ω m)
Graphite	2.267	8.517	119–165	1–2	10^{-4}
Diamond	3.515	6.155	900–2300	10	10^{11}–10^{18}

1.4 Bonding energy and energy bands

We can wonder what the result of the interaction of two single orbital atoms is, for instance for hydrogen atoms. Considering the spatial exponential decay of the 1s orbital, it is safe to say that there is no interaction when the atoms are distant. The symmetry of the atoms also implies that the electrons are on the same energy level. When the distance between atoms approaches the decay radius of 1 Å, the orbitals begin to overlap each other, allowing the electrons to transfer from one atom to the other. This charge redistribution thus creates new energy levels. Denoting this coupling energy t12 and the energy of the original 1s orbital Es, we can calculate the bonding state energy level 2Es–2t12, and the anti-bonding state one 2Es+2t12. A covalent bond is therefore formed when the two electrons are in the lower energy bonding level. This is the traditional model used by chemists to represent atomic bonding in molecules. It can be generalized to other types of orbitals and structures. The Homo level of a chemical compound is the highest occupied energy level, while the Lumo level is the lowest unoccupied energy level (figure 1.9).

This model can also be generalized to a crystalline structure composed of a substantial amount of periodically arranged hydrogen atoms. New energy levels will be created by coupling the orbitals of close atoms. The number of interactions between orbitals increases with the number of atoms in the crystal, therefore increasing the splitting of energy levels, even creating a quasi-continuum called the energy band instead of discrete energy levels when the number of atoms is high enough.

The energy band dispersion E(k) of an infinite 3D periodic crystal with one orbital per atom in a cubic lattice can be obtained with the tight binding method:

$$E(\vec{k}) = \epsilon_0 - 2|t|(\cos k_x a + \cos k_y b + \cos k_z c) \quad (1.1)$$

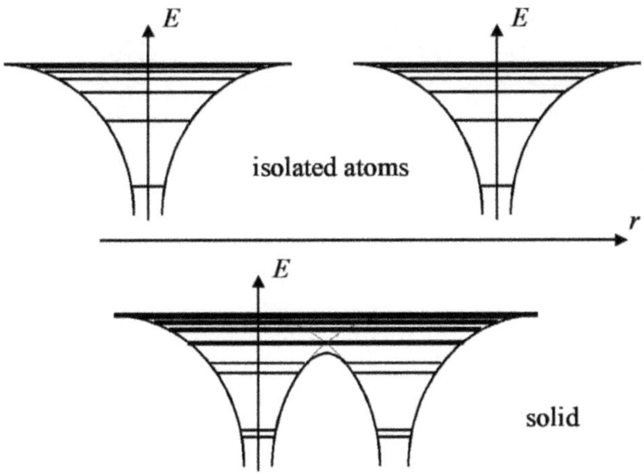

Figure 1.9. Schematic description of the energy spectrum change from single atoms to a solid. Every discrete energy level splits into two separate energy levels when the atoms are bound in a solid. Reprinted from [4] by permission from Springer Nature, copyright 2019.

$$E(\vec{k}) = \epsilon_0 - 6|t|\left(1 - \frac{k^2a^2}{2}\right) \text{ when } k \to 0 \tag{1.2}$$

$$\text{where } \frac{1}{m^*} = \frac{1}{\hbar^2}\frac{d^2E}{d^2k_x} \text{ so that } \frac{1}{m^*} = \frac{2\,|t|\,a^2}{\hbar^2} \tag{1.3}$$

where a is the lattice constant, t the bonding energy, **k** the wave vector, m^* the effective mass and ε_0 the orbital level.

Equation (1.1) reveals that, indeed, the energy dispersion forms a continuum for an infinite periodic crystal, as the splitting of energy levels is infinitely small. If we neglect electronic correlations, an electric field would easily set in motion electrons as there would not be a large energy barrier to surmount, leading to a metallic behavior. This is not what happens in insulators or semiconductors, where a common electric field cannot move charges across the energy gap. The effective mass, defined in equation (1.3), takes into account these electronic correlations, and can be interpreted as the efficiency with which an applied electric field would accelerate an electron in the crystal based on Newton's law (see [4] or standard quality textbooks in solid-state physics listed below).

Coupling different energy shells can generate energy bands, but this concept is still valid for one-orbital atoms, usually when the unit cell is composed of two or more atoms.

To illustrate this, we can model the situation by considering a linear periodic chain of atoms, were the unit cell contains two different atoms, labelled A and B. They both contain only one orbital, the energies of which are EA and EB (figure 1.10). Denoting the coupling energy t, a tight binding calculation gives rise to two energy bands instead of one:

Figure 1.10. Linear chain of two different atoms. Reprinted from [4] by permission from Springer Nature, copyright 2019.

$$2E(k) = (E_A + E_B) \pm [(E_A - E_B)^2 + 8t^2(1 + cos2ka)]^{1/2}. \tag{1.4}$$

The lowest orbital energy (EA) forms the lower band, while the highest orbital energy (EB) forms the upper band: an electron from an atom A must go through the higher energy of the B orbital to reach the next equivalent site. If initially only the atoms A have electrons, the Pauli principle states that the lower band should be completely filled while the upper band should be completely empty at a temperature of 0 K. This is a semiconductor with a band gap of $2E_g = [(E_A - E_B)^2 + 16t^2]^{1/2}$.

This simple argument can be extended to two-dimensional and three-dimensional lattices: at least two atoms per (periodic) unit cell are required to form a band gap. For silicon and germanium, for instance, the fcc Bravais lattice includes two atoms per unit cell, and explains the famous semiconductor properties [4].

Based on the precision required, band structures can be computed in various ways. The simplest and least precise methods are tight binding, linear combination of atomic orbitals and nearly free electron calculations. In some cases, such as transition metals or layered compounds, more exact methods must be employed. Density functional and pseudo-potential methods are thus called '*ab initio*' methods as they make as few assumptions as possible about the crystal and manage to compute accurate electronic and phononic structures. The lattice structure defines an array of atomic potentials used to solve the Schrödinger equation for the outermost valence electrons. An effective core due to the deepest shells and ions is used to represent the scattering potential of the atoms. A plane wave basis is used to compute the Schrödinger equation, while electron–electron interactions are modelled by a self-consistent electric field based on the Hartree–Fock theory [10]. The most common way to solve Schrödinger's equation is to use the density functional method (DFT) to calculate the electron density rather than determine the wave function [10–13].

1.5 Summary

In this chapter, we looked at the basic properties of the carbon atom and analyzed its ability to create chemical bonding to form various compounds and allotropes. This survey is by no means complete as this field is constantly evolving. The following chapters will be devoted to studying each one of these allotropes, to analyze their electronic properties and their applications. Synthesizing and preparing these materials require technical solutions the reader can look for in the numerous and detailed literature on the subject. We introduced here the concept of electronic bonding in periodic systems, leading to the Bloch energy band structure theory and the related new physical concepts, such as holes in valence bands, or effective mass of electrons and holes. This basis should help the reader to understand the concepts dealt with in the next chapters, but it is strongly recommended to examine further these basic concepts within the specialized literature.

1.6 Exercises

(Q1) Describe the different carbon bonding arrangements, and for each one provide a material example. In your opinion, in what field can organic technology outperform inorganic one?

(Q2) Explain the mechanisms behind sp^2 and sp^3 hybridization, and describe hybridization in silicon, in germanium and in III–V compounds. Hybridization notion doesn't appear when computing the band structure using an *ab initio* calculation, explain why.

(Q3) Describe the method to obtain the bonding energy between atoms from the atomic orbital energy of each one of them.

(Q4) Explain Hund's rules.

References and further reading

[1] Demarchi D and Tagliaferro A 2014 *Carbon for Sensing Devices* (Berlin: Springer)

[2] Falkowski P, Scholes. R J, Boyle E, Canadell J, Canfield D, Elser J, Gruber N and Hibbard K *et al* 2000 The global carbon cycle: a test of our knowledge of earth as a system *Science* **290** 291–6

[3] Bruckner R *Advanced Organic Chemistry* (Amsterdam: Elsevier)

[4] Razeghi M 2009 *Fundamentals of Solid State Engineering* 3rd edn (Berlin: Springer)

[5] Audi G, Wapstra A H, Thibault C, Blachot J and Bersillon O 2003 The NUBASE evaluation of nuclear and decay properties *Nucl. Phys. A* **729** 3–128

[6] de Laeter R, Böhlke J K, De Bièvre P, Hidaka H, Peiser H S, Rosman K J R and Taylor P D P 2003 Atomic weights of the elements. Review 2000 (IUPAC Technical Report) *Pure Appl. Chem.* **75** 683–800

[7] Wieser M E 2006 Atomic weights of the elements 2005 (IUPAC Technical Report) *Pure Appl. Chem.* **78** 2051–66

[8] Cottrell T L 1958 *The Strengths of Chemical Bonds* 2nd edn (London: Butterworths)

[9] Karthik P S, Himaja A L and Singh S P 2014 Carbon allotropes synthesis methods applications and future perspectives *Carbon Lett.* **15** 219–37

[10] Madelung O 1978 *Introduction to Solid State Theory* (Berlin: Springer)

[11] Ashcroft N W and Mermin D N 1976 *Solid-state physics* (New York: Holt, Rinehart and Winston)

[12] Ziman J 1964 *An Introduction to Solid State Physics* (Cambridge: Cambridge University Press)

[13] Parr R G and Yang W 1994 *Density Functional Theory of Atoms and Molecules* (Oxford: Oxford University Press)

Chapter 2

Diamond

2.1 Introduction

One of the most beautiful solids in nature is the tetrahedrally bonded three-dimensional crystal called diamond, which is made of pure carbon. In this chapter we review the basic properties of diamond and examine the band structure of face-centered cubic (fcc) diamond. We observe that pure diamond is a perfect insulator, but one of its outstanding features is that it can be boron doped, which makes it conduct metallically and then be used to make transistors. Just as with Si and Ge, diamond is an indirect bandgap material, which means that it does not luminesce well. Luminescence can be recovered by engineering defects around which the indirect gap symmetry is broken. What makes Si so useful is the self-passivation property via its own oxide layer. With diamond, hydrogen can be used to efficiently passivate the surface dangling bonds. It turns out that the boron-doped hole conduction remarkably gives rise to superconductivity at very low temperatures ~5 K. The superconductivity is believed to be of the normal Bardeen–Cooper–Schrieffer (BCS) type where singlet pairs are formed via electron-phonon-induced attraction. With nitrogen vacancy (NV) controllable defect engineering, it is possible to make coupled coherent spin pairs which have long enough coherence times. The quantum coherence survives long enough to be useful for quantum mechanical storage and processing of information. Spin information can be stored and retrieved using suitably polarized laser pulses.

2.2 Introduction to diamond

Let us start this chapter by studying one of the most fascinating and beautiful materials on earth, pure tetrahedrally bonded carbon, namely diamond [1–3].

Diamond consists of pure carbon connected via sp3 in a tetrahedral bonding arrangement, as shown in figure 2.1. It is a transparent, hard, semiconducting material with a large bandgap which exhibits some remarkable properties, some of which will be described below. It is normally found in the natural state, but it can

doi:10.1088/2053-2563/ab35d1ch2

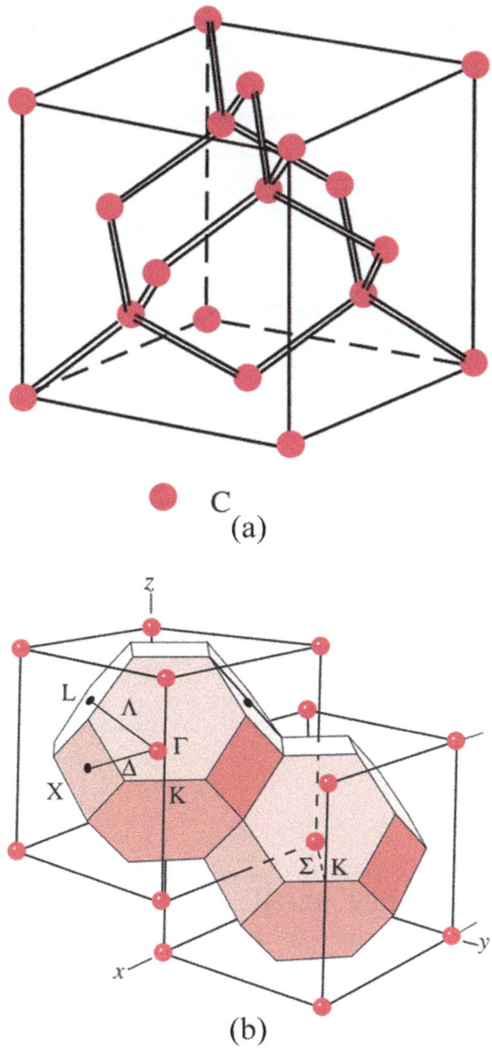

Figure 2.1. (a) Diamond structure. (b) First Brillouin zone (BZ) of an fcc lattice (Wigner–Seitz cell). Reprinted from [2] by permission from Springer Nature, copyright 2010.

also be produced in small quantities, thin films, by subjecting carbon allotropes to extremely high pressure. Graphite is indeed combined with a metal catalyst and heated to temperatures higher than 1800 K, under high pressures (3–5 GPa). This growth method is the called the high-pressure high-temperature (HPHT) method [4]. Let us examine lattice structure and electronic band structure.

2.3 Band structure of diamond

The lattice structure of diamond is shown in figure 2.1(a).

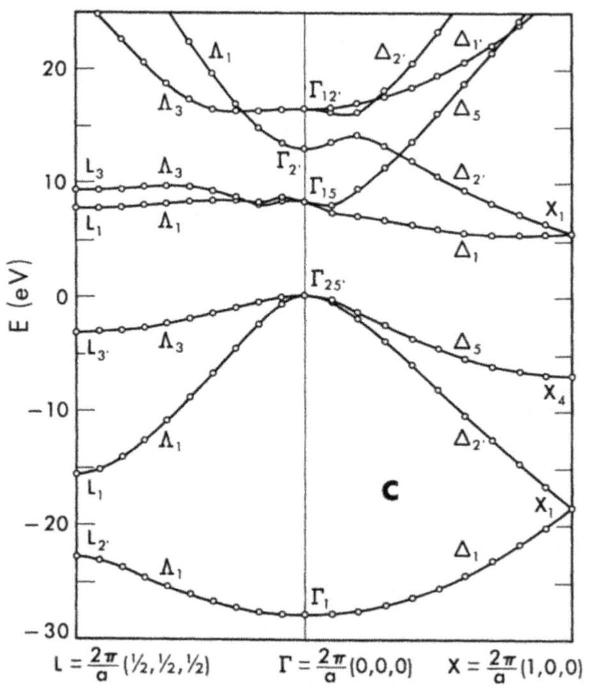

Figure 2.2. Band structure of diamond. Reprinted with permission from [5]. Copyright 1966 by the American Society.

The Bravais lattice is fcc. The basis consists of two identical atoms displaced from each other by a quarter of the cube body diagonal. The Brillouin zone (BZ) together with the nomenclature for the points inside the BZ are shown in figure 2.1(b).

The band structure of diamond evaluated with the pseudo-potential method is shown below.

The electronic parameters given by this computation are as follows:

E_g = 5.46–5.6 eV indirect gap (shown in figure 2.2).

$E_{\Gamma 1}$ = 7.3–7.4 eV direct gap (shown in figure 2.2).

E_{so} = 6 meV spin–orbit splitting energy.

2.3.1 Some properties of the diamond electronic structure

Just as with Si and Ge, diamond is an indirect bandgap semiconductor with electron effective masses of $\boldsymbol{m_l = 1.4\ m_o}$ and $\boldsymbol{m_t = 0.36\ m_o}$. The valence holes' effective masses are $\boldsymbol{m_h = 2.12\ m_o}$ for heavy holes, $\boldsymbol{m_{lh} = 0.7\ m_o}$ for light holes and $\boldsymbol{m_{so} = 1.06\ m_o}$ for the split-off band. Diamond is also an excellent electrical insulator and has the highest breakdown electric field of any known material. Diamond is the best known naturally occurring thermal conductor. Diamond is highly transparent. Synthetic nanocrystalline diamond is the hardest material known. See also [1, 2, 6].

2.3.2 Doped diamond and applications

Natural blue diamonds are semiconductors due to substitutional boron impurities replacing carbon atoms. Diamond is a good electrical insulator, having a resistivity of 100 GΩ · m to 1 EΩ · m (10^{11} to 10^{18} Ω · m). Diamond can be doped. A particularly interesting scenario occurs when boron is used as the dopant. This case is discussed below in more detail. Doped diamond can become conductive and has important applications in electronics; see, for example, the work of Garrido *et al* [7] published in the review book.

The in-plane transistor shown in figures 2.3 and 2.4 is made using hydrogen surface passivated boron doped diamond and is a particularly good example of modern diamond electronics. The traditional applications of diamond are related to its superb thermal conductivity and its hardness. More recently it has also become a material for electro-optic applications. Remarkably, NV color centers in diamond, which are defects, can be made reproducibly (see figure 2.5). These defects exhibit controllable spin coherence on a scale of single spins with very long coherence time. This spin coherence, it is claimed, can form the basis for the hardware of quantum information processing and computers.

The NV defect center in diamond is an exceptionally versatile single-spin system with unique quantum properties that have driven its application in diverse areas ranging from quantum information and photonics to quantum metrology. Cryogenic scanning magnetometry stands out as potentially the most impactful application of NV centers, taking advantage of the exquisite magnetic field sensitivity and intrinsic atomic scale of the NV center for high-resolution imaging. The operation of an NV-based magnetic probe is dependent on a fundamentally

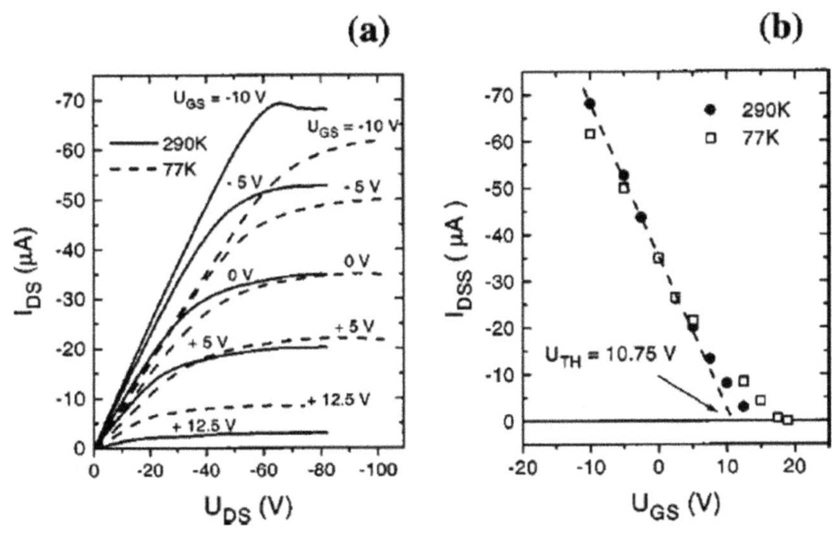

Figure 2.3. Transistor characteristics (290 and 77 K) of a device with 300 nm channel width. (a) Drain–source current (I_{DS}) versus drain–source voltage (U_{DS}) and different gate voltages (U_{GS}); (b) saturation current (I_{DSS}) versus U_{GS}. Reprinted from [7], with permission of AIP Publishing.

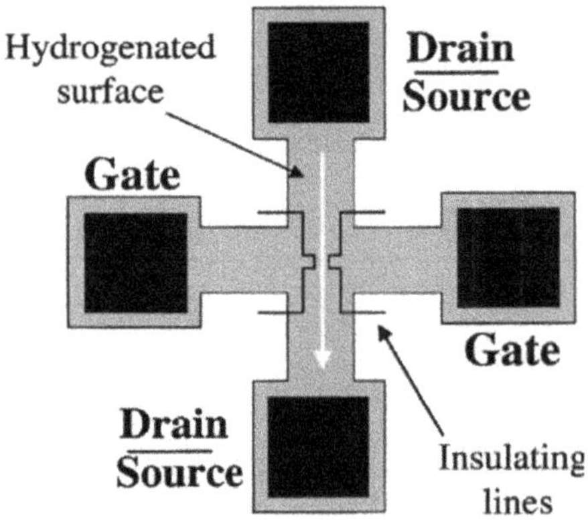

Figure 2.4. In-plane gate transistor. Reprinted from [7], with permission of AIP Publishing.

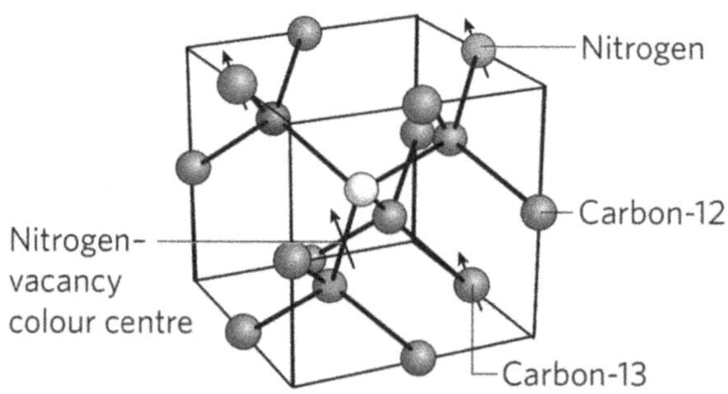

Figure 2.5. Nitrogen vacancy (NV) colour center defect in diamond. Reprinted from [11] by permission from Springer Nature, copyright 2008. See also [12].

different sensing principle than other imaging methods, namely the spin-dependent photoluminescence of a solid-state defect. NV centers maintain their high field sensitivity over a large temperature range—from cryogenic to ambient and above—and hence are ideal for imaging nanoscale magnetism through orders of magnitude in temperature. A cryogenic NV scanning magnetometer enables the study of a host of new systems that could benefit from a highly sensitive nanoscale probe, particularly solid-state systems with non-trivial magnetic order or magnetic phase transitions at low temperature [8].

2.4 Superconductivity in p-doped diamond

Boron has one less electron than carbon and, because of its small atomic radius, boron is relatively easily incorporated into diamonds. As boron acts as an acceptor, the resulting diamond is effectively p-doped and conducts via holes in the valence band. The HPHT method can be used to produce heavily doped material with carrier concentrations exceeding 10^{20} cm^{-3}. However, this material shows a limited conductivity due to the compensation of carriers by nitrogen, so

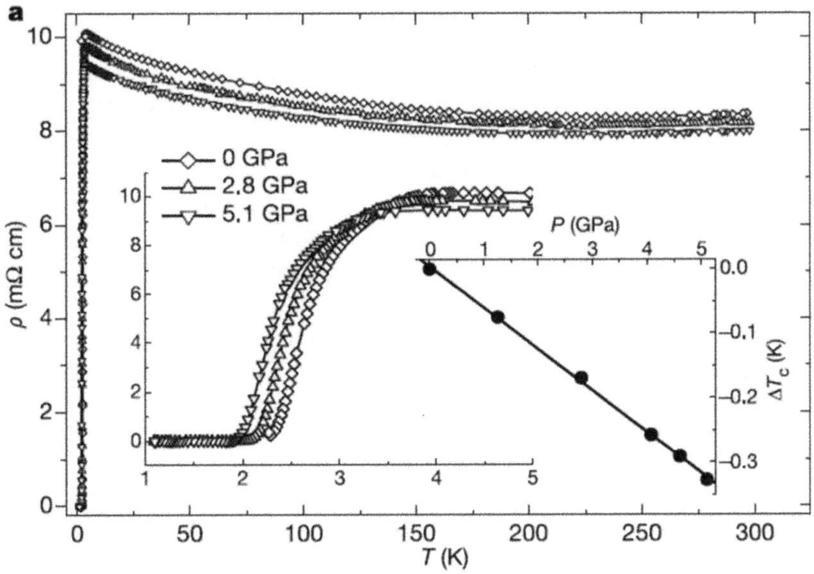

Figure 2.6. Electrical resistivity for B-doped diamond at normal and representative high pressure. The inset shows the details of the resistivity behavior below 5 K and the pressure induced shift ΔT_c of the midpoint of the resistive transition. Reprinted from [10] by permission from Springer Nature, copyright 2004.

HPHT is actually limiting the applications of B-doped diamond. Chemical vapor deposition (CVD) is now the most versatile growth technique to produce this material. Methane is mixed with hydrogen and diborane under plasma, then deposited on a substrate where diamond nanoparticles were nucleated. The substrate temperature is usually around 1200 K (the plasma is 2500 K), and the reactor pressure is maintained at 25 mbar. Traditionally used substrates (Si, Nb, W) can withstand these growth conditions, but their thermal expansion induces intrinsic residual stress that can be reduced by carefully designed growth conditions. The result is a polycrystalline material that shows a roughness varying with the growth duration, from large grains and rough surfaces due to protruding crystallites, to nanocrystalline B-doped diamond with nanometer-scale roughness. A careful polishing can reduce the roughness without affecting the electrochemical properties of the material [4]. One of the most exciting aspects of B-doped diamond is that at low enough temperatures (T = 4 K) it becomes a superconductor (figure 2.6).

B-doped diamond up to 10^{21} cm^{-3} is a Type II superconductor below 4 K. The superconductivity survives up to fields of B = 3.5 T. The mechanism for superconductivity has not been investigated in great detail. The superconductivity is believed to be of the normal BCS type where singlet pairs are formed via electron-phonon-induced attraction. More work needs to be done to prove or disprove the mechanisms, especially in view of the similarity of diamond to materials such as doped Si and Ge. Boron-doped Si above a solubility limit of several per cent has recently been shown to undergo a superconductive transition at $T_c = 0.35$ K [9].

2.5 Summary

Pure carbon forms a tetrahedrally bonded crystal called diamond which has, on top of its traditional applications, recently been discovered to have unique quantum properties. In this chapter we looked at the band structure and some of the outstanding properties such as boron doping which makes it conduct metallically and be used to make transistors. Just as with Si and Ge, diamond is an indirect bandgap material which means that it does not luminesce well. What makes Si so useful is the self-passivation property via the oxide layer. With diamond, hydrogen can be used to efficiently passivate the surface dangling bonds. The boron-doped hole conduction remarkably gives rise to superconductivity at very low temperatures ~5 K. The superconductivity is believed to be of the normal Bardeen–Cooper–Schrieffer (BCS) type where singlet pairs are formed via electron-phonon-induced attraction [9, 10]. With nitrogen vacancy (NV) controllable defect engineering, it is possible to make coupled coherent spin pairs which have long enough coherence time to be useful for quantum mechanical processing of information. Spin information can be stored and retrieved using suitably polarized laser pulses (see [11, 12]).

2.6 Exercises

(Q1) (a) How would you calculate the energy levels of a single electron on a hydrogen molecule using the free electron method, and then benzene? Outline the method and explain the difficulties, if any.
(b) What happens when another electron is added?

(Q2) Explain what is meant by thermal conductivity. Why is diamond such a good thermal conductor, and what do you think amorphous carbon is like? Do some research on amorphous carbon and compare its properties to diamond.

(Q3) Explain, mainly in words and with diagrams, how a field effect transistor (FET) works.

(Q4) Impure diamond is often coloured: can you explain with some research where the various colours come from ? Could one change the colour by ion implantation?

References and further reading

[1] Pan L S and Kania D R (ed) 1995 *Diamond: Electronic Properties and Applications* (Kluwer)
Pan L S and Kania D R (ed) 1994 *Diamond: Electronic Properties and Applications* (Berlin: Springer)

[2] Yu P Y and Cardona M (ed) 2010 *Fundamentals of Semiconductors: Physics and Materials Properties* (Berlin: Springer)

[3] Walker J 1979 Optical absorption and luminescence in diamond *Rep. Prog. Phys.* **42** 1605–59

[4] Cobb S J, Ayres Z J and Macpherson J V 2018 Boron doped diamond: a designer electrode material for the twenty-first century *Annu. Rev. Anal. Chem.* **11** 463–84

[5] Saslow W, Bergstresser T K and Cohen M L 1966 *Phys. Rev. Lett.* **16** 355

[6] Razeghi M 2009 *Fundamentals of Solid State Engineering* 3rd edn (Berlin: Springer)
[7] Garrido J A, Nebel C, Todt R, Rosel G, Amann M and Stutzmann M 2003 Fabrication of in-plane gate transistor on hydrogented diamond surfaces *Appl. Phys. Lett.* **82** 988
[8] Pelliccione M, Jenkins A, Ovartchaiyapong P, Reetz C, Emmanouilidou E, Ni N and Bleszynski Jayich A C 2016 Scanned probe imaging of nanoscale magnetism at cryogenic temperatures with a single-spin quantum sensor *Nat. Nanotechnol.* **11** 700–5
[9] Bustarret E *et al* 2006 Superconductivity in doped cubic silicon *Nature* **444** 465
[10] Ekimov E, Sidorov V, Bauer E, Melnik N, Curro N, Thompson J D and Stishov S 2004 Superconductivity in diamond *Nature* **428** 542
[11] Hanson R and Awschalom D 2008 Coherent manipulation of single spins in semiconductors *Nature* **453** 1043
[12] Knowles H S, Kara D M and Atatuere M 2014 Observing bulk diamond spin coherence in high purity nanodiamonds *Nat. Mater.* **13** 21
[13] Vogel D, Krueger P and Pollmann J 1997 Tight binding method in semiconductors *Phys. Rev.* B **55** 12836
[14] Zhang W, Ristein J and Ley L 2008 Hydrogen-terminated diamond electrodes. II. Redox activity *Phys. Rev.* E **78** 041603

Chapter 3

Carbon fullerenes

3.1 Carbon fullerene: an introduction

This chapter is devoted to exploring the properties of the unique and beautiful molecule called buckminsterfullerene C60, often referred to as 'Buckyballs'. Buckyballs are brilliant examples of the power and ingenuity of organic chemistry and chemists. We describe here the energy levels of the molecules and the band structure of the crystalline solid form of C60 in the fcc lattice. The doped structures are reviewed, and the remarkable high-T_c superconductivity of the potassium-doped fullerene K_3C_{60} is put into focus. It turns out that doping with longer molecular side-groups can increase the lattice constant and that this, contrary to intuition, actually raises the superconducting transition temperature T_c. This astonishing observation, we believe, gives researchers clues as to what the mechanism for pairing might be. One may here discover mechanisms which go beyond the standard Bardeen–Cooper–Schrieffer (BCS) theory of electron–phonon-induced superconductivity. Doping fullerenes can also give ferromagnets and insulators, and the fullerene molecule, which has strong acceptor properties, has found great applications in polymer solar cell technology. This is treated in chapter 6.

3.2 Carbon fullerenes: the fullerene molecule

The theoretical prediction of a stable carbon closed-cage molecular structure dates back to 1970, but the experimental demonstration happened 15 years later, at which point existence of fullerenes was verified [1]. At the time carbon clusters were produced by laser ablation of graphite. The resulting molecules were divided into two categories based on a mass spectra analysis: 10–30 atom rings, and larger molecules with 60- and 70-member atoms. Researchers predicted a closed-cage configuration for these high-mass molecules. They gave it the name 'fullerene' after R Buckminster Fuller, a famous architect building geodesic domes that looked like these newly discovered molecules (figure 3.1).

doi:10.1088/2053-2563/ab35d1ch3

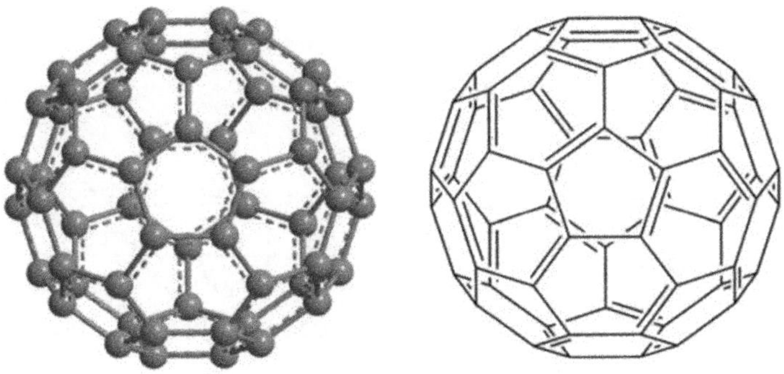

Figure 3.1. Two schematic views of the C60 fullerene molecule. Figure reused from [2], reprinted by permission of Taylor & Francis Ltd.

The fullerene lattice and the hexagonal graphite one are very similar as they are both a two-dimensional surface. However, curving a graphene sheet requires substituting pentagons to some hexagons. Geometrically speaking, polyhedral theory requires exactly 12 pentagonal faces of the closed-case arrangement, and any additional number of hexagonal faces to create multiple arrangements that form a closed structure. Twelve pentagons therefore compose the smallest possible fullerene, C20. The out-of-plane bending of the sp^2 bonds result in an extra strain energy that can be reduced by separating the pentagons in the lattice as much as possible. The smallest structure in which all pentagons are isolated is C60. Because of its structural stability, C60 is the most abundant product of any fullerene growth process, usually 3–6 times more abundant than the second most, C70. Usually referred to as 'Buckyballs', the C60 molecules present a soccer-ball arrangement composed of 12 pentagonal and 20 hexagonal faces. Carbon–carbon bonds are either single bonds of 1.46 Å along the 60 pentagonal edges, or double bonds of 1.40 Å between adjacent hexagons [1, 3, 4]. Nuclear magnetic resonance (NMR) measurement of the molecular diameter is on average 7.10 Å [3, 4], close to the geometrical diameter of 7.09 Å from the model considering the atoms as points. The icosahedral symmetry of C60 can be used to approximate its electronic structure. Each carbon atom contributes to four valence electrons. The σ-bonded sp^2 electrons can be safely neglected as core-level molecular states, while the 60 radially oriented p_z orbital electrons form the valence states [5]. The molecular orbital eigenfunctions are determined thanks to the irreducible representations of the icosahedral point group. They can be approximated through the spherical harmonics $Y_{l,m}$ as the spherical shape of Buckyballs would suggest. The spherical shape of C60 suggests an approximation of these molecular orbitals based on spherical harmonics. Each angular momentum state l can host 2(2l + 1) electrons, so the first 50 electrons fill all states up to $l = 4$, leaving 10 electrons for the 22 $l = 5$ states. Many computational methods can be used to calculate the relative energy splittings within each level. The lowest energy = 5 states belong to the 5-fold degenerate h_u representation. These

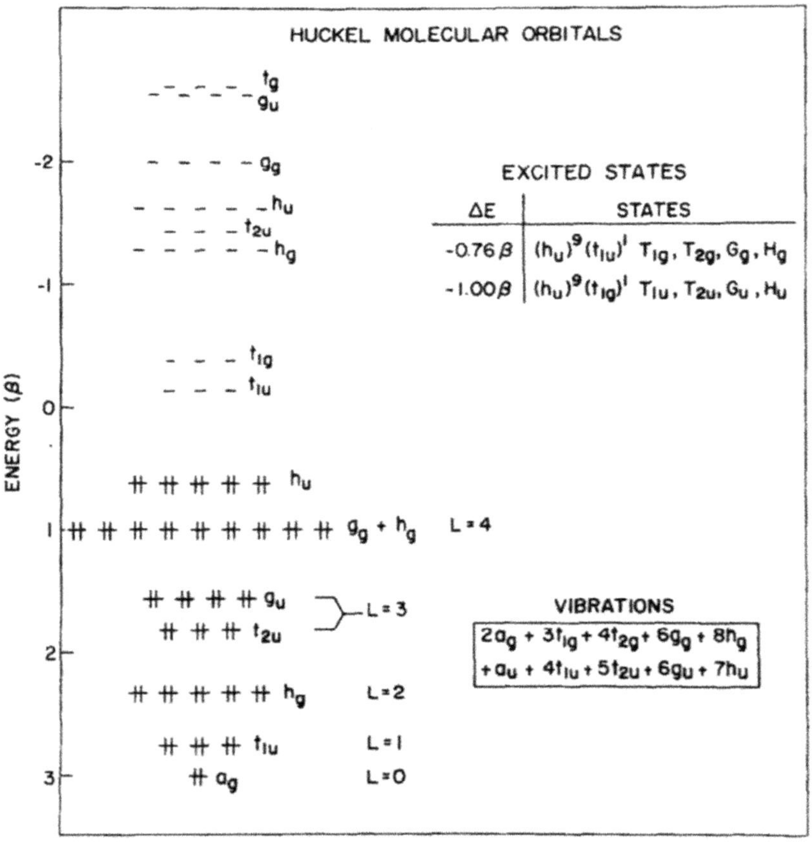

Figure 3.2. Energy-level diagram of Hueckel π-molecular orbitals for C_{60}, ground state. From [6], reprinted with permission of Elsevier.

states stand for the remaining 10 electrons and constitute the highest occupied molecular orbitals (HOMOs) in the ground state.

The lowest unoccupied molecular orbitals (LUMOs) are t_{1u} states (figure 3.2), experimentally observed 1.9 eV above the h_u levels [3, 5, 7]. Being three-fold degenerate, and transforming into one another under rotations around the [111] axis, the t_{1u} molecular orbitals are similar to p atomic orbitals. The h_u orbitals have also transformation properties similar to d atomic orbitals. Bulk C60 is at room temperature a molecular solid with a fcc structure, held together by van der Waals interactions. While the nearest-neighbor distance is 10.02 Å, the intermolecular separation (2.92 Å) is close to the inter-layer spacing in graphite (3.35 Å). Van der Waals interactions are not as strong as covalent bonds, so molecules appear to rotate freely at room temperature. However, below 260 K rotations freeze, while molecules orient themselves relative to one another, leading to a simple cubic crystal structure. The solid's electronic structure shows bands derived from the molecular orbitals of the individual molecules. The undoped solid has a semiconductor behavior with a

1.5 eV band gap between the hu-derived valence band and the t_{1u}-derived conduction band, which have narrow bandwidths (0.4 eV) [3, 7]. It is possible to dope C60 solids with alkali metal, alkali earth, or other elements, therefore changing significantly the material conduction properties, even allowing superconductivity in some cases [7]. This is discussed in the next section.

Looking at the energy levels in figure 3.2 we note that the 'spherical' symmetry leads to high-level degeneracy degrees with a band gap of about 1.8 eV. Thus, the first LUMO level has a 3-fold degeneracy and HOMO has 5-fold degeneracy, which is lifted by structural relaxation [7].

3.3 Useful molecules derived from fullerenes

One of the most useful fullerene derivatives is PCBM, shown in figure 3.3. It has the strong acceptor properties of a fullerene ball and is used in conjunction with fast charge generation via photoexcitation of bound excitons. The PCBM acceptor is connected through a molecular chain to a donor head. When photoexcitation occurs, the exciton splits with the positive hole transferring to the donor and the electron to the acceptor where they can exist for a long time; see chapter 6 where this is discussed in more detail.

In general, fullerenes embrace a wide variety of potential shapes, symmetries and electronic configurations, of which some are systematically preferred according to experimental observation [8]. Structure bond analysis is done by Rogers *et al* [9] by considering leapfrog fullerenes constructed by omnicapping and dualising a fullerene parent.

3.4 Electron-doped fullerene superconductivity

The first fullerene molecule to be discovered, and the family's namesake, buckminsterfullerene (C60), was prepared in 1985 by Richard Smalley, Robert

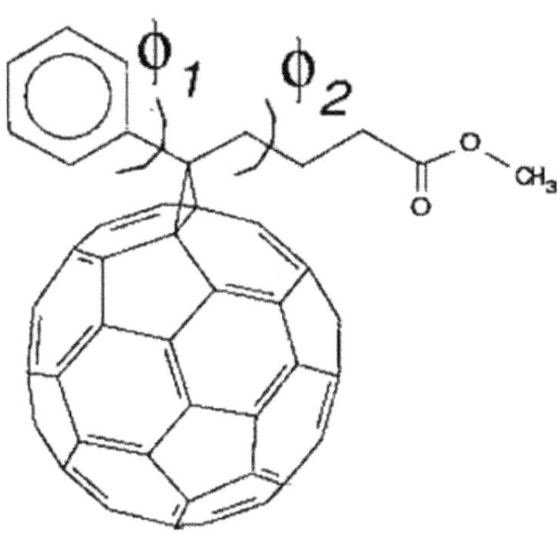

Figure 3.3. Structure of PCBM, a fullerene derivative, showing definitions of dihedral angles φ_1 and φ_2. From [10], reprinted with permission of the American Chemical Society.

Curl, James Heath, Sean O'Brien, and Harold Kroto at Rice University. Kroto, Curl, and Smalley were awarded the 1996 Nobel Prize in Chemistry for their roles in the discovery of this class of molecules. It was Haddon and colleagues [11] who discovered that the fullerenes could be doped with alkali metal and made conductive. A liquid–liquid interface precipitation method is generally used to synthesize carbon fullerene-based materials. The band structure of undoped fcc fullerene is shown in figure 3.4 just below. The fermi energy is defined by the zero line.

Subbands around the fermi energy for solid C60 can be divided in two: the bands at about −0.5 eV are the h_u bands which are occupied in solid C60 and the bands at about 1.5 eV are the t_{1u} bands which become populated in AnC60 [7, 12] (see figure 3.5).

In K_3C_{60} superconductivity sets in at $T = 40$ K, which is an extremely high temperature for any material, particularly an organic one with such narrow energy bands. Researchers have shown that here too it is possible to derive an electron–phonon-induced pairing force as done in the BCS theory. A detailed phonon

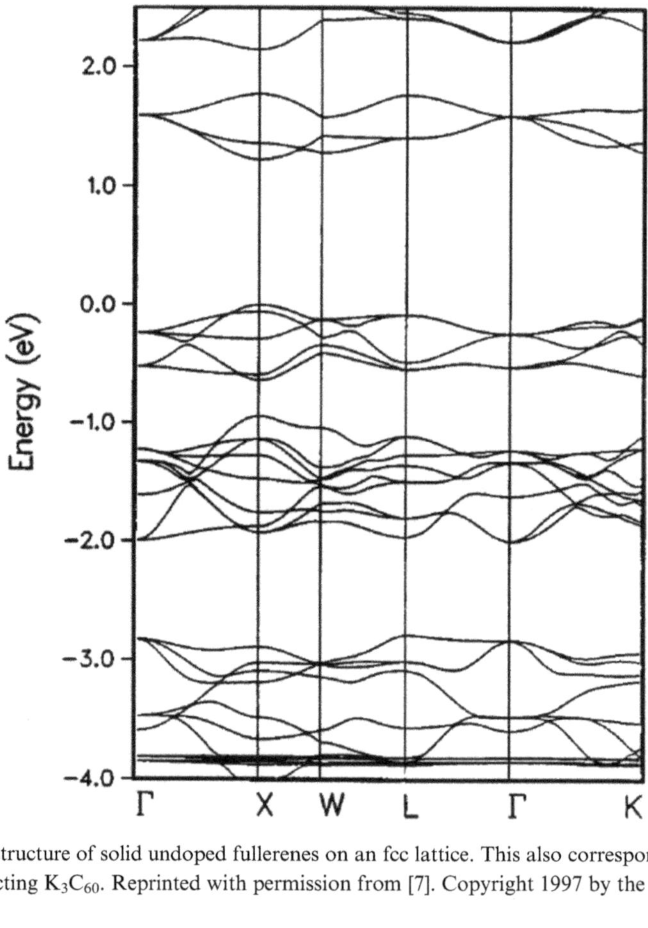

Figure 3.4. Band structure of solid undoped fullerenes on an fcc lattice. This also corresponds to the structure of the superconducting K_3C_{60}. Reprinted with permission from [7]. Copyright 1997 by the American Physical Society.

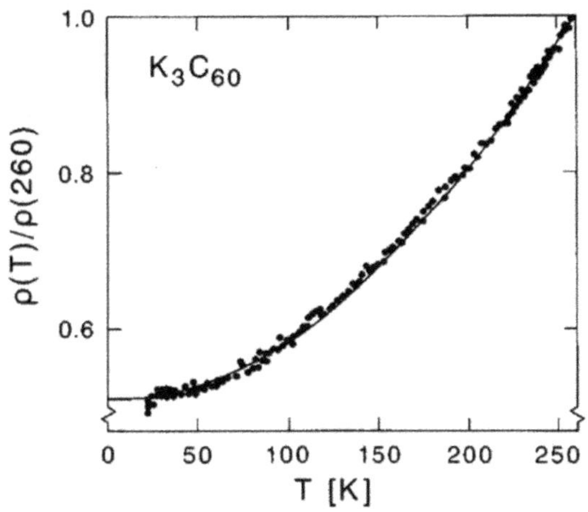

Figure 3.5. *T*-dependent resistivity $\rho(T)$ of K_3C_{60} normalized to the value at T = 260 K dotted curve experimental, solid curve theoretical. Reprinted with permission from [7]. Copyright 1997 by the American Physical Society.

spectrum and coupling analysis is given in Gunnarsson's paper and can indeed give BCS-type attraction. However, in contrast to a simple metal, the doped fullerene metal is highly correlated, meaning that when two electrons occupy the same ball orbital, there is a strong Coulomb repulsion, called Hubbard U, which is of ~1.5 eV and therefore much larger than both bandwidth and the bare electron–phonon attraction. The naïve BCS argument therefore does not work, and this is not the only situation where the simple BCS picture breaks down. The problem is even more pronounced in the newer ceramic high-T_c materials [13, 14]. Theoreticians have therefore tried to come up with other mechanisms [14] or modified BCS mechanisms that would explain the superconductivity in strongly correlated metals. The most commonly accepted mechanism for fullerenes is the delayed electron–phonon attraction mechanism proposed in various ways by various authors [7, 15]. The argument here is that a phonon-induced coupling change takes time to come about. This is because the first electron arrives on the ball and moves the ions, and the ions are slower than electrons: this then polarizes the medium and generates the attraction for the second electron which arrives on the ball. The time delay implies that when the second electron arrives, it sees a net attraction because the first electron which caused the polarization has moved away already and the strong Coulomb repulsion has been reduced. In other words, this works when the electron bandwidth is much larger than the phonon energy. A renormalized Coulomb repulsion then becomes

$$\mu^* = \frac{\mu}{1 + \mu \ln(B/\omega_{ph})}, \tag{3.1}$$

where μ is the Coulomb repulsion, B = 1/2 times the electronic bandwidth, and ω_{ph} the dominant mode in the coupling. This type of explanation, though popular and

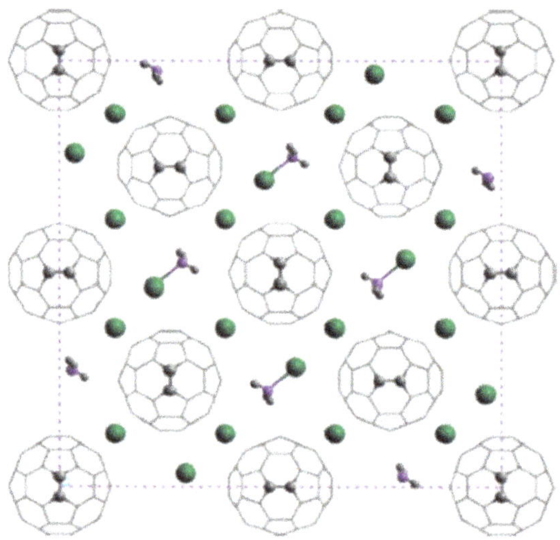

Figure 3.6. Schematic representation of a planar projection of the crystal structure of $NH_3K_3C_{60}$. Dots are potassium atoms; other dots surrounded by three gray dots are NH_3 molecules. Reprinted with permission from [16]. Copyright 2009 by the American Physical Society.

highly plausible, has however not satisfied everyone in the community and the search for new mechanisms is far from over. Other alternative models have been proposed, dealing with expanded fullerenes, for instance [16] (figure 3.6).

Finally, we note that though we have emphasized the fascinating superconductivity of fullerene complexes, there are a multitude of other exciting effects, such as ferromagnetism, associated with other complexes of C_{60} [3]. The role of Hund's rule coupling is a result of the 3-fold degeneracy of the t_{1u} orbital to which three electrons are transferred from the dopant [16].

In figure 3.7, the superconducting and antiferromagnetic (insulating I) order parameter Δ is plotted by Capone *et al* as a function of the ratio of the Hubbard correlation energy U (on ball repulsion) and the bandwidth W. This ratio changes (increases) as the lattice constant is increased. We see that, remarkably, the Tc is helped by electronic correlations up to a point beyond which the insulating antiferromagnetic order takes over. Capone *et al* assume a strong correlation model and show that they can reproduce the experiment far better than modified BCS can [16].

3.4.1 Digression: high-T_c cuprate superconductors

One of the first materials in the class of high-T_c superconductors is abbreviated as YBaCuO. The unit cell is shown in figure 3.7(b). The Tc is ~93K—a great breakthrough when it was discovered by Mueller and Bednorz [13]. Even though it is a little outside the subject matter of this book, it is thought by many researchers [17] that the next breakthrough to room temperature Tc is likely to come from carbon compounds. So, the reader can join in with this exciting speculation having read the material here.

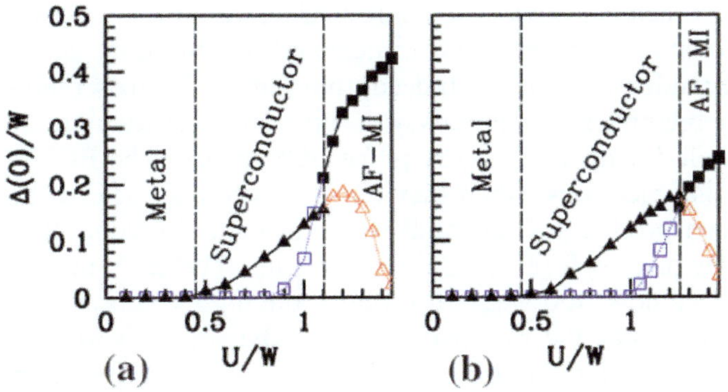

Figure 3.7(a). Superconducting solution (triangles) and antiferromagnetic solution (squares) derived using dynamical mean field theory [16]. The first-order transition between the two phases is indicated by a vertical line separating the superconductor from the antiferromagnetic Mott insulator. The case in the presence of a frustrating next neighbor hopping $t' = 0.3t$ absent in (a). Reprinted with permission from [16]. Copyright 2009 by the American Physical Society.

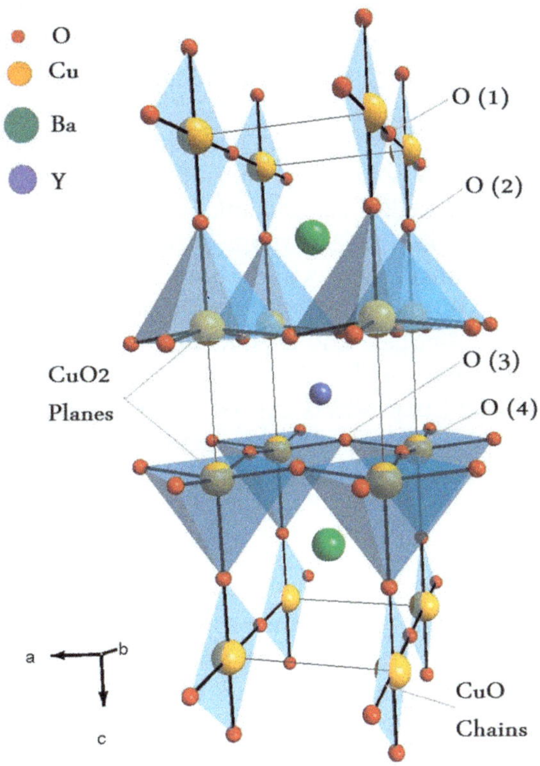

Figure 3.7(b). YBaCuO unit cell. This image has been obtained by the author from the Wikimedia website, where it is stated to have been released into the public domain. It is included within this book on that basis.

In YBaCuO, which has one of the clearest structures, it is believed that the pair of CuO planes shown in the diagram are critical and constitute the smallest unit which produces superconductive pairing. Indeed, thin films will superconduct if these two CuO planes are present. For materials with superconductivity up to about 30 K, the accepted and well-documented theory is the BCS theory. The BCS theory is based on the notion that electrons, when they move in a lattice, deform the lattice. This deformation then produces a potential which attracts a second electron and thus allows a Cooper pair to form at low temperature. The BCS mechanism is, most researchers believe now, too weak to produce the high-T_c observed with cuprates, and it is suggested that the explanation has to be sought with electron–electron correlations; in particular, building on the ideas of P W Anderson [13] and the notion that a moving spin (not just charge) will tend to attract to itself spins of the opposite direction and this will have a similar 'water bed' effect as with phonons. A complete explanation with quantitative explanation is still lacking and the subject is still hotly debated.

3.4.2 Fullerene nanowhiskers

The fullerene nanowhiskers (FNWs) terminology represents all needle-like crystals comprising fullerene molecules with diameters less than 1000 nm. The words 'nanorod' and 'nanowire' are here replaced with 'nanowhiskers' to avoid confusion, as was described [18, 19].

Takeya *et al* synthesized FNWs and intercalated the fibers with potassium; see also the review in [19]. FNWs are thin crystalline fibers of fullerene molecules including C60 and C70 endohedral or functionalized fullerenes. The FNWs display n-type doping and are used in a diverse range of applications including field effect transistors, solar cells, chemical sensors and photocatalysts (figure 3.8).

Alkali-doped C60 NWs have realized the highest superconducting volume fraction of the alkali-metal-doped C60 crystals and display a high critical current density under a high magnetic field of 50 kOe (figure 3.9).

Fullerene nanofibers and nanosheets can be synthesized using the liquid–liquid interfacial precipitation (LLIP method). This method relies on diffusion of a poor solvent of fullerenes such as isopropyl alcohol into a fullerene-saturated solution [18–20].

3.5 Summary

Chapter 3 was devoted to the sensational properties of the unique and beautiful molecule known as buckminsterfullerene: an astonishing example of what organic chemistry can achieve. The energy levels of the molecules and the associated crystalline solids in fcc were described. The amazing doped structures were reviewed, and the remarkable high-T_c superconductivity of the potassium-doped fullerene K_3C_{60} was put in focus. The fact that increasing the lattice constant actually raises T_c in a given range of side-group-induced stretch has given researchers clues as what the mechanism for pairing might be, in contrast to standard Bardeen–Cooper–Schrieffer (BCS) theory. However, it should be mentioned that narrowing the band

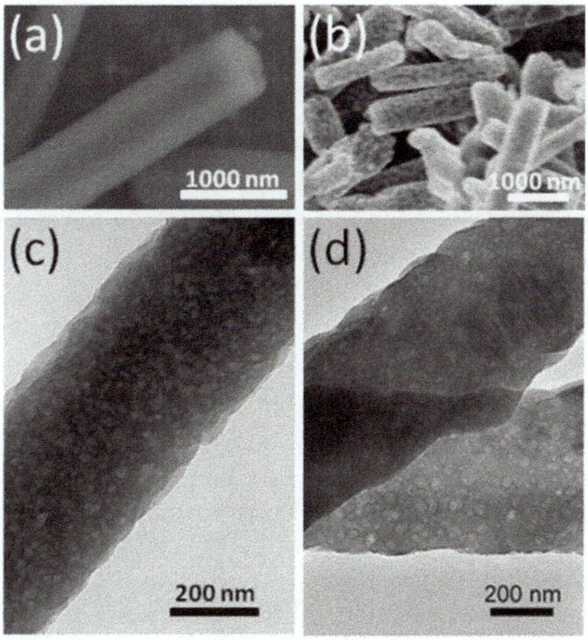

Figure 3.8. SEM micrographs of (a) C_{60} NW and (b) K-doped C_{60} NWs. (c) and (d) are the micrograph of (a) and (b) observed using a transmission electron microscope. From [19], licensed under a Creative Commons Attribution License (CC BY 3.0).

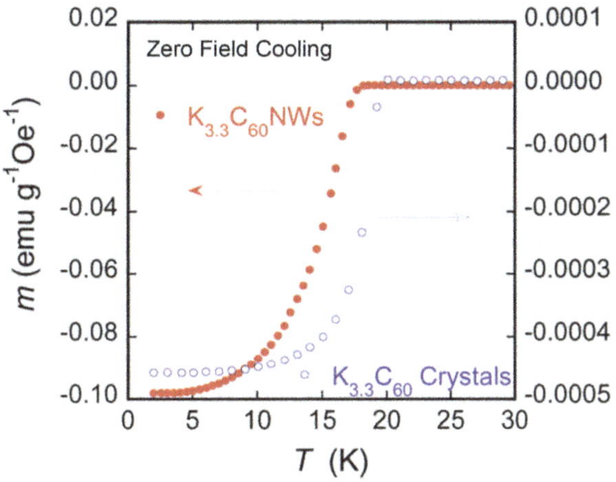

Figure 3.9. The temperature dependences of the normalized magnetic moment m for the $K_{3.3}C_{60}$ NWs and $K_{3.3}C_{60}$ samples. The left and right ordinates of the graph set separately for $K_{3.3}C_{60}$NWs and $K_{3.3}C_{60}$ crystals are the magnetic moment normalized by the applied magnetic field and the weight of the samples. From [19], licensed under a Creative Commons Attribution License (CC BY 3.0).

increases the density of states at the Fermi level and this raises T_c in the framework of BCS, as well. But doping fullerenes can also create ferromagnets [21] and insulators, and the fullerene molecule has strong acceptor properties which have found great applications in polymer solar cell technology. This is treated in chapter

6. More recently, Japanese researchers managed to synthesize FNWs with diameters less than 1000 nm and potassium dope the fibers by intercalation. In this way they were able to make superconducting wires with transition temperatures reaching 17 K. This material combines lightweight plasticity with superconductive properties with a respectable T_c of 17 K. It reached the highest superconducting volume fraction in this class of materials. Fullerenes have many other exciting applications as organic semiconductors in the field of field effect transistors and photoconductors; we have here focused on the truly novel and sensational superconductive properties.

3.6 Exercises

(Q1) Explain the difference between Bose and Fermi statistics. Give the formulae and explain what happen in the two cases as more particles are added at $T = 0$ K. What is Bose–Einstein condensation?

(Q2) When two electrons are added to the LUMO level of C_{60}, do you expect an electron–electron repulsion U to occur in the same orbital? Find out how big it is estimated to be, do some research.

(Q3) Explain the origin of Hund's rule coupling in atoms and how it could apply to electron-doped fullerenes. Remember that the empty, undoped orbitals are degenerate?

References and further reading

[1] Kroto H W *et al* 1985 C_{60}: buckminsterfullerene *Nature* **318** 162–3

[2] Liu Q *et al* 2014 The applications of buckminsterfullerene C60 and derivatives in orthopaedic research *Connect. Tissue Res.* **55** 71–9

[3] Forro L and Mihaly L 2001 Electronic properties of doped fullerenes *Rep. Prog. Phys.* **64** 649

[4] Haddon R C 1992 Electronic structure, conductivity and superconductivity of alkali metal doped (C60) *Acc. Chem. Res.* **25** 127

[5] Hornbaker D J 2000 Electronic structure of carbon nanotubes systems measured with scanning tunneling microscopy *PhD Thesis* University of Illinois at Urbana Champaign

[6] Haddon R C, Brus L E and Krishnan R 1986 Electronic structure and bonding in icosahedral C60 *Chem. Phys. Lett.* **125** 459–64

[7] Gunnarsson O 1997 Superconductivty in fullerides *Rev. Mod. Phys.* **69** 575

[8] Fowler P W and Manolopoulos D E 1995 *An Atlas of Fullerenes* (Oxford: Oxford University Press)

[9] Rogers K M and Fowler P W 2001 Leapfrog fullerenes Hueckel bonding order and Kekule Structures *J. Chem. Soc., Perkin Trans.* **2** 18–22

[10] Cheung D and Troisi A 2010 Theoretical study of the organic photovoltaic electron acceptor PCBM: Morphology, electronic structure, and charge localization *J. Phys. Chem. C* **144** 20479–88

[11] Haddon R C *et al* Conducting films of C_{60} and C_{70} by alkali-metal doping 1991 *Nature* **350** 320

[12] Erwin S C and Pickett W E 1992 band structure of K_3C_{60} *Phys. Rev.* B **46** 14257

[13] Bednorz J G and Müller K A 1986 Possible high T_c superconductivity in the Ba-La-Cu-O system *Zeitschrift für Physik B.* **64** 189–93
[14] Anderson P 1987 The resonating valence bond state in La_2CuO_4 and superconductivity *Science* **235** 1196–8
[15] Bogolioubov N N, Tolmachev V and Shirkov D V 1958 *A New Method in Superconductivity* (New York: Consultants Bureau)
[16] Capone M *et al* 2009 Modeling the unconventional superconducting properties of expanded A_3C_{60} *Rev. Mod. Phys.* **81** 943
[17] Razeghi M private communications
[18] Miyazawa K 2015 Synthesis of fullerene nanowhiskers using the liquid –liquid interfacial precipitation method and their mechanical electrical and superconducting properties *Sci. Technol. Adv. Mater.* **16** 013502
[19] Takeya H 2012 Superconducting fullerene nanowhiskers *Molecules* **17** 4851–9
[20] Kuwasaki M K *et al* 2002 C_{60} nanowhiskers formed by the liquid–liquid interfacial precipitation method *J. Mater. Res.* **17** 83
[21] Allemand P M *et al* 1991 TDAE doped fullerene *Science* **253** 301
[22] Lieber C M and Chen C-C 1994 Preparation of fullerene and fullerene based materials *Solid State Phys.* **48** 109
[23] Martin N, Solladie N and Nierengarten J F 2006 Advances in molecular and supramolecular fullerene chemistry *Electrochem. Soc. Interface Summer* 29
[24] Bardeen J, Cooper L and Schrieffer J R 1957 BCS theory *Phys. Rev.* B **106** 162
Bardeen J, Cooper L and Schrieffer J R 1957 BCS theory *Phys. Rev.* B **108** 1175
[25] Erwin S and Pederson M 1993 Electronic structure of superconducting Ba_6C_{60} *Phys. Rev.* B **47** 14657(R)
[26] Xiang X, Hou J, Briceno G, Vareka W, Mostovoy R, Zettel A, Crespi V and Cohen M L *et al* 1992 Synthesis and electronic transport of single crystal K_3C_{60} *Science* **256** 1190

IOP Publishing

The Mystery of Carbon
An introduction to carbon materials
Manijeh Razeghi

Chapter 4

Graphene

4.1 Introduction

In this chapter we review some of the basic properties of the recently discovered one-atom-thick but robust wonder material called graphene. Graphene can be grown on a substrate or it can be studied in suspended form. We present the explanations of where its new properties come from and try to point out why there is so much excitement about the prospects (and now reality) of using graphene in high technology. The material in this chapter covers band structure, the band structure associated novelty, and some applications. We also review the properties of graphene nanoribbons and bilayers. Having studied and identified its novelty, we may then ask: what can we learn from graphene-like band structures? Is it worth trying to reproduce them artificially in other materials? What more can we learn about many-body effects using the rather more simple configuration of graphene? This is perhaps one of the most interesting questions.

In 2004 Novoselov, Geim and coworkers at the University of Manchester used a simple mechanical exfoliation technique to obtain supported single-layer graphene [1–3]. Graphene was indeed first made by applying a scotch tape to graphite and removing a single monolayer of hexagonally bonded carbon. Since then researchers have succeeded in making graphene in a variety of ways, and it has generated an enormous interest. Its truly astonishing features are extreme mechanical robustness combined with good tunable conductivity. The combination of high optical transparency (98%) and high metallic conductivity is already enough to make this one of the most interesting and useful materials. Let us study the band structure and discover more.

Graphene synthesis

The traditional and most effective way of making two-dimensional sheets of graphene is mechanical exfoliation, providing the best quality possible for its synthesis. However, this method has a poor yield and requires complicated experimental skills which limit its applications. Chemical exfoliation is an

doi:10.1088/2053-2563/ab35d1ch4

alternative that has a great yield but is limited by defects that can be introduced by the process. Chemical vapor deposition is the traditional method to grow two-dimensional material, and the quality, size and thickness can be tuned by the experimental conditions (temperature, pressure, flows). A cost-effective alternative can be found with electrochemical deposition, producing thin films of nanomaterials at low temperatures. However, this technique is still limited by the quality and the irreproducibility of its results. Finally, massive production of thickness-controlled layers can be achieved through solution phase synthesis.

4.2 Graphene band structure

4.2.1 Tight-binding method

The tight-binding method is the easiest and most common way of computing graphene band structure. A valence orbital can be assigned to every carbon atom, and by coupling them we can generate energy bands. In the following, we denote A and B the two atoms of the graphene lattice (see figure 4.1), and we consider the tight-binding Hamiltonian neglecting the Coulomb interactions between electrons. $c_{i\sigma}{}^{+}$ and $c_{i\sigma}$ are the creation and annihilation operators for electrons at atomic orbital σ and a given site i [2]

$$H_e = \sum_{i,\sigma} \varepsilon_{i\sigma} c_{i\sigma}^{+} c_{i\sigma} + \sum_{i \neq j,\sigma} t_{ij} c_{i\sigma}^{+} c_{j\sigma} \tag{4.1}$$

We note $\varepsilon_{i\sigma}$ the atomic orbital energies with spin σ, and t the tight-binding coupling matrix element linking two neighboring orbitals.

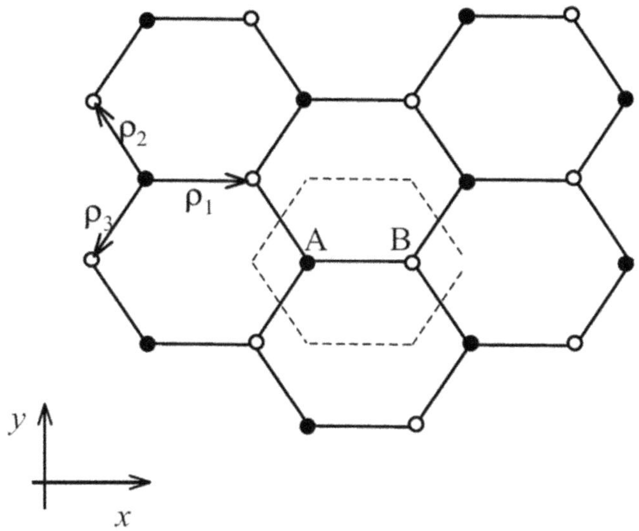

Figure 4.1. Hexagonal structural parameters of graphene. Atoms A and B are both carbon but topologically inequivalent. Reprinted with permission from [1].

Using this Hamiltonian in the Heisenberg equation [3, 4] for a type-A atom, for example, we have (noting E the energy)

$$(E - \varepsilon_{i\sigma})c_{i\sigma}^{a} = \sum_{j} t_{ij} c_{j\sigma}^{b}. \tag{4.2}$$

The sum j goes over the n.n. There is an asymmetry between A- and B-type atoms which is not equivalent by translational symmetry, even if they seem physically equivalent, so the Bloch theorem cannot be directly used. We also have a similar relation for the B-type atoms:

$$(E - \varepsilon_{i\sigma})c_{i\sigma}^{b} = \sum_{j} t_{ij} c_{j\sigma}^{a}. \tag{4.3}$$

Now we can substitute equation (4.3) in equation (4.2) and relate two equivalent A or B atoms at distance R by the Bloch's phase factor exp $(ik.R)$ in the usual way and solve the equations. We find there is an unusual linear dispersion at the so-called Dirac points in k space K and K′, where there is no energy gap. This is because of the mirror symmetry of the A and B atoms, and not a translational symmetry. A new pseudo-spin quantum number emerges from this topological restriction. This notion can be generalized to other topologies, which could have far more complex chiral symmetries. We can have degenerate energies, 'hole-like' levels and 'particle-like' ones. One only needs to consider the myriad of complex topologies that can be found in artists' designs, in churches, on walls, on cloth tiles, ceramics, etc. See figure 4.2, where the different points are nearly always logically equivalent, but not topologically identical, until one goes a very long distance. In the language of Bloch band theory, one would have multitudes of closely packed energy bands giving rise to pseudo-angular momentum quantum numbers, which for the dual space of

Figure 4.2. Jam Mosque Yazd (Iran). Note the beautiful topology and compare with the background blue substrate graphene. Note the rotational symmetry of the inner pattern. What is the density of states and the carrier mobility on this lattice which could be made artificially with nanocrystals? This is an exciting challenge!

graphene is like a spin. The topological constraints can be considered analogous to a pseudo-relativistic constraint, and the solutions can acquire new spin or fractional spin-like quantum numbers.

There is no reason to believe that this is only restricted to physics. A large number of mathematical problems in all sorts of areas of human knowledge could have this type of 'pseudo-spin' solution structure (in the Dirac formalism spinor-like solutions are necessary because of linearization), not just the abstract tight-binding connection of patterns. It will be interesting to discover some such overlap and connectivity in biology, botany, economics, or synergetics and information science. What it means here is that we have situations in which there are delocalized eigenstates which can have both particle-like and hole-like character and are degenerate. Trying to form a localized 'wavepacket' implies that one has to mix particle with hole states. The particle can, so to speak, slip into a close-by hole, and then reappear somewhere else maybe a long way away in space as a particle, just as a particle in relativistic quantum theory (see p. 130 of [3]), where it is argued that at very short times, the fabric of space time is rich and more complex, allowing electrons to make excursions into holes and reappear as the particle component of the pair. Thus, extraordinarily, spin is a feature which is not lost by going to non-relativistic average speeds (lower than light speed), actual particle velocity being light velocity as shown by Dirac. The spin concept survives and becomes the intrinsic property of an electron. Short enough times, or simply time evolution, imply large energy uncertainties. This inevitable energy uncertainty, which is a consequence of time evolution, would normally imply that the particle can have higher than light 'instantaneous' velocities, but this is not allowed by relativity, so instead it generates a new dimension: particle anti-particle space, and the spin quantum number.

Coming back to graphene, the problem however is that true Bloch symmetry is a very special situation, and in reality there is always some disorder which removes these exact mathematical features. So, if one wants to claim the new science, one must also consider the slightly disordered version of the graphene Hamiltonian. In a semiconductor such as Si, for example, Bloch periodicity is not the only reason for the existence of energy gaps and effective masses. The short-range bonding order and bond couplings are enough to partially recreate these features, and this is true also for many of the novel features in materials such as, for instance, nanocrystals, nanodots, wires, etc. Also, the Dirac electron spin survives because excursions into the anti-particle world are always possible but is this also true for the new pseudo-spin dimension [2] (figure 4.3).

The first thing to note about disorder is the appearance of a finite density of states at the fermi energy which will immediately affect the transport properties. The effect of disorder in energy regions where the density of state is small implies Anderson localization, which in turn implies that the average spatial distance to a close-by energy level is long. Thus, disorder at band edge gives rise to Anderson localized states [3, 5]. The recent simulation work of Chandrasekhar and Bush has gone some way towards explaining why a honeycomb topology is better than a square lattice for quantum transport in the presence of the same degree of disorder [6]. This is a significant step forward and demystifies the advantages of graphene. It also points

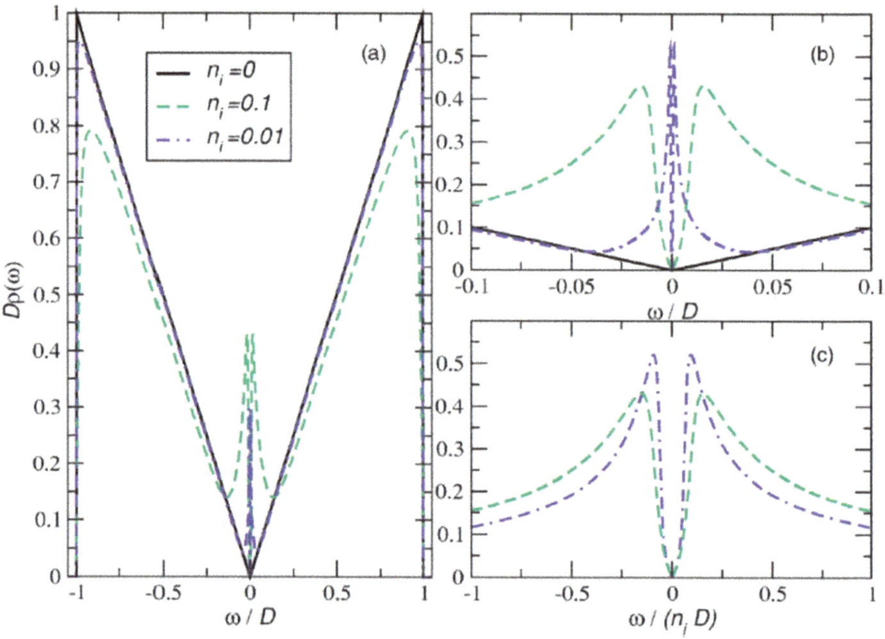

Figure 4.3. Density of states as a function of energy. Density of states ρ at the fermi level multiplied by the cut off energy D in units of the cut off energy D for different values of the impurity density n_i. (a) for the entire range, (b) ρ for $\omega \ll D$, (c) ρ as a function of ω/n_iD indicating an energy scale of the order of $n_i\ D/4$. Reprinted with permission from [3]. Copyright 2006 by the American Physical Society.

the way to finding even better topologies for possible nanotechnology applications. Apart from topology, the size of the systems is also crucial, because it has been shown rigorously that in a two-dimensional network, any amount of disorder causes the localization of all eigenstates in an infinite lattice. Anderson localization enhances the importance of electronic interactions, and electron–phonon interactions, because the spatial presence of the particle is less diffuse and more concentrated: they see each other more when close and they deform the lattice more effectively. Perhaps the most interesting and immediate effect of disorder in graphene is the **self-doping** that occurs with charge transferring from the (disorder-produced) slightly higher energy to the slightly lower regions. This is expected around the defects, and a result of weak screening. This should even be more pronounced in lattice with Penrose tiling if they could be made to carry charge (see the topology in figure 4.2 and [4]). The metallic or diffusive character of transport would normally survive and may even increase with weak disorder because it enhances the density of states at the fermi level. Importantly, the screening remains ineffective even though the electronic structure is (gapless) and in this sense a 'metal', so that electronic correlations are more pronounced, charged defects and charge pools will occur with weakly screened long-range potentials. The reader is referred to the work of Peres *et al* [5] for a full discussion of what disorder does to graphene, and what one should expect in an imperfect structure or one subjected to potential

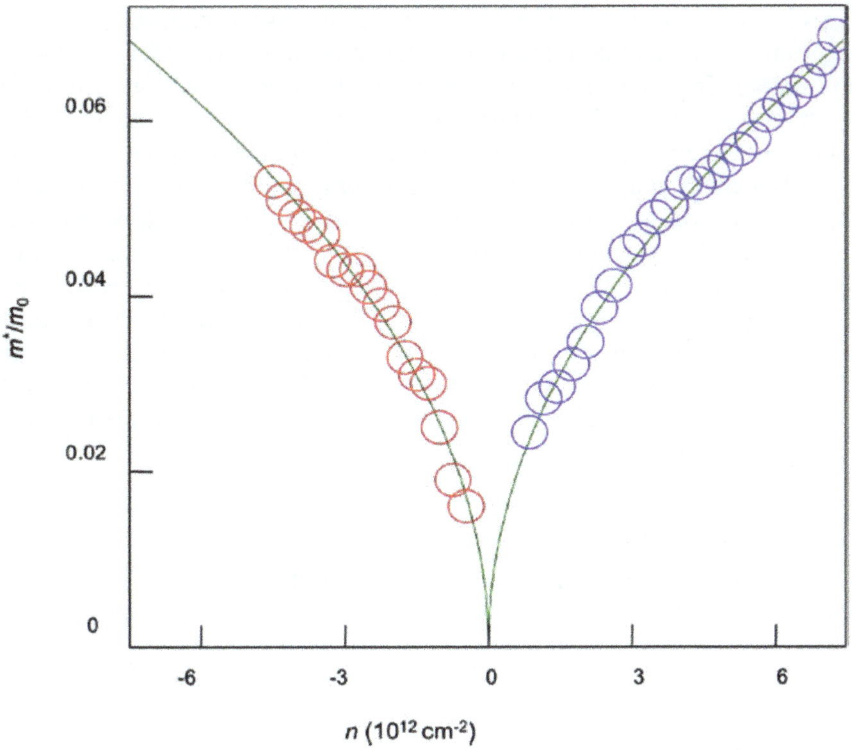

Figure 4.4. Cyclotron mass versus carrier density. Reprinted with permission from [2]. Copyright 2009 by the American Physical Society.

fluctuations. Which one of the 'anomalies' survive and in what form? This is the critical question and maybe one can do newer physics with suitably potential engineered graphene than meets the eye at first. Whether the Dirac-like 'pseudo-spin' concept survives disorder is an interesting question, but beyond the scope of this review. As an example of a truly observed effective mass anomaly, we point to figure 4.4 from Castro *et al* which shows how the cyclotron mass scales with carrier density in graphene. Finally, and importantly, scientists have recently fabricated bilayer graphene sheets with a small twist in the topology (angle of 1%–2%) so that the carbon atoms are no longer exactly on top of each other as in bilayer graphene, but there is now an overlayer-induced superlattice structure. The effect of the coupling in the superlattice topology is to form new bands with lower periodic symmetry. Applying a gate voltage, it is possible to inject charges into this new system where the authors have demonstrated superconductivity transitions at critical twist angles with T_c up to 1.7 K. The new physics is fascinating since it proves that this superconductor is not of the phonon-induced type but indeed of Mott–Anderson type with band filling just off the Mott insulating state.

The cyclotron mass is a measure of how the carrier semiclassically responds to the Lorentz force in a magnetic field from which one can infer how it responds to any force. This is true as long as we reason in the semiclassical picture. There is also a fully quantum mechanical derivation of an 'effective mass' valid for any finite band quantum system. In particular, for a tight-binding system even with disorder, an effective mass was derived by S Datta [7] using the density matrix linearized in a force field. Datta used this generalized definition of 'effective mass' to determine the sign of the Hall effect in a disordered system. In graphene, the high mobility and the low effective mass will have to survive weak disorder if they are to be acceptable as novel properties. Indeed, as mentioned already, it has been shown that the hexagonal nanoribbon topology, for example, suppresses the scattering from impurities compared to a square topology [6]: hexagonal structures are protected against backscattering. But we shall see later that high mobility at high temperature in graphene is a result of the high energy of optic phonons, which unlike in GaAs 2DEGs, do not act at normal temperatures to scatter the charge. Here topological considerations do not suffice.

4.2.2 Graphene wavefunction

Another way to derive the band structure (figure 4.5) is to use the real spatial wavefunction ψ and then the Wallace expectation value and optimization process [8]

$$\psi_{\mathbf{k}}(\mathbf{r}) = \sum_{\mathrm{A}} \exp(i\mathbf{k}\cdot\mathbf{R}_{\mathrm{A}})X(\mathbf{r}-\mathbf{R}_{\mathrm{A}}) + \lambda\sum_{\mathrm{B}} \exp(i\mathbf{k}\cdot\mathbf{R}_{\mathrm{B}})X(\mathbf{r}-\mathbf{R}_{\mathrm{B}}), \tag{4.4}$$

$\mathbf{r}$ is the particle coordinate, $\mathbf{R}_{\mathrm{A}}$ and $\mathbf{R}_{\mathrm{B}}$ are position of the two types of atoms see figure 4.6 and $\mathbf{k}$ is the Bloch wavevector, $\lambda = 1$ or -1:

$$t_i = \int X*(r-\mathbf{R}_{\mathrm{A}})HX(r-\mathbf{R}_{\mathrm{B},i})d\mathbf{r}, \quad E_{\mathbf{k}} = E_0 \pm \left|\sum_i t_i \exp(-i\mathbf{k}\cdot\boldsymbol{\rho}_i)\right|. \tag{4.5}$$

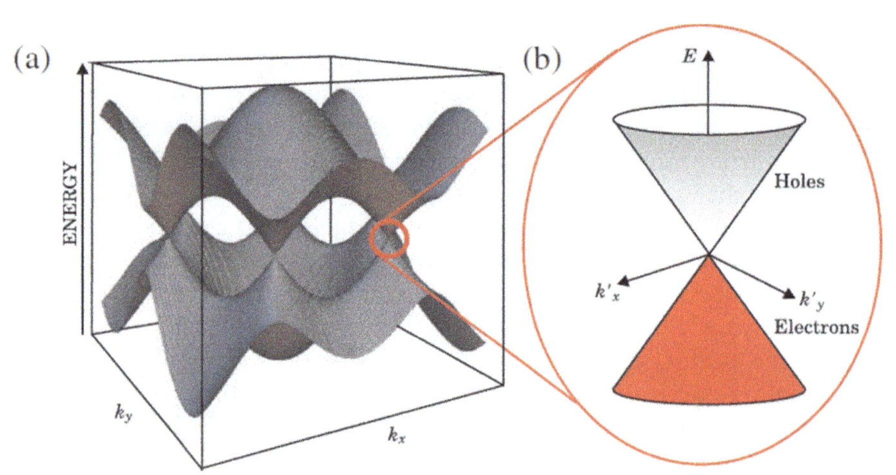

Figure 4.5. Band structure of graphene plotted in three dimensions to exhibit the zero gap or Dirac points K, K′. Courtesy of Professor Philip Kim.

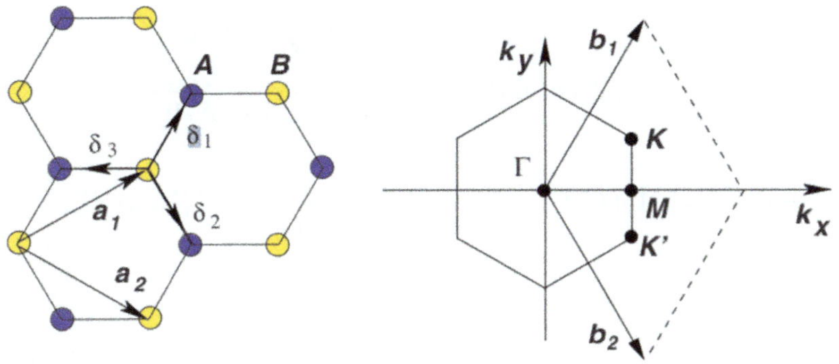

Figure 4.6. Graphene lattice and first Brillouin zone. Near the K, K′ points, shown above, now called Dirac points, the dispersion (energy momentum relation) is linear implying a zero effective mass. Reprinted with permission from [2]. Copyright 2009 by the American Physical Society.

An analytic tight-binding band structure computation gives

$$E(k) = \pm\gamma_0\sqrt{1{+}4\cos\left(\frac{3}{2}k_x a\right)\cos\left(\frac{\sqrt{3}}{2}k_y a\right) + 4\cos^2\left(\frac{\sqrt{3}}{2}k_y a\right)} \tag{4.6}$$

where γ_0 is the bonding energy t.

The honeycomb structure can be thought of as a triangular lattice with a basis of two atoms per unit cell with two-dimensional lattice vectors $a_1 = (a/2)(3, \sqrt{3})$ and $a_2 = (a/2)(3, -\sqrt{3})$, $a = 0.142$ nm is the carbon–carbon distance $K = (2\pi/(3a), 2\pi/(3\sqrt{3}\ a))$ and $K' = (2\pi/(3a), -2\pi/(3\sqrt{3})\ a)$ as the inequivalent corners of the first Brillouin zone and are called 'Dirac points'. The Dirac points play a role similar to the role of Γ points in direct band gap semiconductors.

Relative to the Dirac point, the dispersion equation becomes

$$E_{\pm}(q) = \pm\hbar v_F k \tag{4.7}$$

with k the norm of the wavevector. The dispersion depends on the fermi velocity v_F. In tight-binding v_F can be expressed in terms of the nearest neighbor hopping integral t so that

$$\begin{aligned} \hbar v_F &= 3ta/2 \\ a = 0.14\ \text{nm},\quad t &= 2.5\ \text{eV},\quad v_F = 10^8\ \text{cm s}^{-1}. \end{aligned} \tag{4.8}$$

4.2.3 Kane theory in the −2-band model with valence and conduction band

The Bloch periodicity changes the free electron band dispersion, which can be computed rigorously near $k = 0$ using k-p theory, and introduces the concept of the effective mass and shows us how the origin of the effective mass can be understood in terms of some simple properties of the system. We recall that with only one valence band (two in all), the free electron dispersion can be related to the dispersion in a lattice by the expansion [3, 5]

$$E_c(k) - E_0(k) = E_c(0) + \frac{\left|\langle v|\frac{\hbar}{m_0}\vec{k}\cdot\vec{p}\,|c\rangle\right|^2}{\{E_c(k) - E_k\} - E_v(0)} \tag{4.9}$$

$$(E_c(k) - E_0(k) - E_c(0))(E_c(k) - E_k - E_v(0)) = \frac{\hbar^2}{(m_0)^2}\vec{k}^2\,|\langle c|\vec{p}\,|v\rangle|^2$$

$$E_0(k) = \frac{\hbar^2 k^2}{2m_0} \tag{4.10}$$

where p is the momentum operator, E_c and E_v are conduction and valence band edges, m_0 is the bare electron mass, the momentum p matrix element is calculated between the valence $|v\rangle$ and conduction band $|c\rangle$ eigenstates:

$$E_c(k) = E_0(k) + \frac{E_c(0) + E_v(0)}{2} \pm \frac{1}{2}\{(E_c(0) - E_v(0))^2 + 16E_P E_0(k)\}^{1/2} \tag{4.11a}$$

where E_P is

$$E_P = \frac{|\langle c|p_x|v\rangle|^2}{2m_0}. \tag{4.11b}$$

The periodic part of the wavefunction is

$$u_{c,k}(\vec{r}) = u_{c0}(\vec{r}) + \frac{\langle v|\frac{\hbar}{m_0}\vec{k}.\vec{p}\,|c\rangle}{\{(E_c(k) - E_0(k)) - E_v(0)\}}u_{v0}(\vec{r}). \tag{4.11c}$$

Note what happens when the valence to conduction gap tends to zero, i.e. when $E_c = E_v$, then we can define effective mass in the usual way and write

$$\frac{1}{m^*} = \frac{1}{m_0\hbar k}\,|\langle c|p|v\rangle| + \frac{1}{m_0} \tag{4.12}$$

$$m^* = m_0\hbar k/|\langle c|p|v\rangle| \rightarrow k \rightarrow 0. \tag{4.13}$$

From equation (4.11), we note that the energy dispersion becomes linear in k as in graphene, indicating that in this way of defining effective mass, a massless particle is generated as k goes to 0, but in this example, it occurs at $k = 0$ not at $\mathbf{k} = \mathbf{K},\mathbf{K}'$.

4.2.4 Density functional theory versus tight binding in nanoribbons

We showed above how easy it is with n.n, tight-binding method (TBM) to derive the energy dispersion for graphene. This result is widely used and contains all the essential physics. From this, one may erroneously conclude that TBM will work for other topologies just as well. Unfortunately this is not so, and whereas TBM predicts zero gap ribbons for particular widths and armchair structures (see figure 4.14), this is not so using density functional theory (DFT) [9]. In DFT it has been shown that the zero

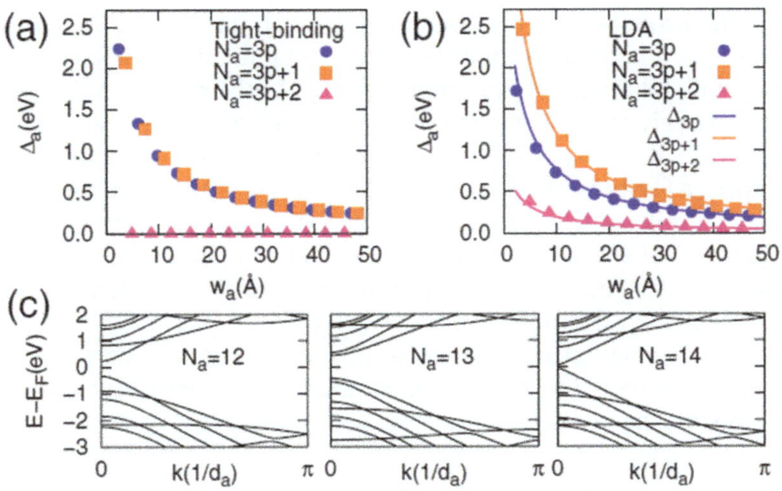

Figure 4.7. The variation of band gaps of N_a-AGNRs as a function of width w_a obtained from (a) TB calculations with a hopping parameter t = 2.70 eV, (b) DFT calculations, (c) DFT band structures of N_a-AGNRs with N_a = 12,13,14 respectvely. N_a = number of dimer lines. GNR = graphene nanoribbon. Reprinted with permission from [9]. Copyright 2006 by the American Physical Society.

gap limit is not reached with nanoribbons, and this has now also been shown experimentally. The exciting physics associated with Dirac points and zero band gaps is therefore in principle to a large extent absent in nanoribbons. But this has to always be looked at in detail, since perfect symmetry is a mathematical artefact, anyway, and significant new physics must be able to survive small perturbations. But this example demonstrates the danger of using approximations which otherwise seem to work perfectly for other ordered carbon structures such as bulk graphene. The finite width and termination edges accentuate the role of exchange and correlations, spin–orbit and further neighbors transfer – all neglected in the simple TBM method.

Young Woo Son *et al* compared the different results computed with the tight-binding method and with DFT. They calculated the band gap of armchair graphene nanoribbons (AGNR) with N_a dimer lines (width of the GNR) (figure 4.7).

4.2.5 Klein tunneling

In a Wentzel–Kramer–Brillouin (WKB) approach [10] the tunneling amplitude T into a high potential barrier region $V(x)$ for a particle with energy E has an exponential decay of the form (m^* = effective mass and $\hbar = h/2\pi$ defined as usual, T_0 the tunnel pre-factor)

$$T = T_0 \exp\left[-\left(\frac{2m^*}{\hbar^2}\right)^{1/2} \int dx'[V(x') - E]^{1/2}\right] \tag{4.14}$$

which implies that if the particle in the barrier has a small mass, tunneling is easier, and indeed in the limit of zero effective mass (linear dispersion) the barrier can be crossed with unit probability, as if it were not there at all. This is apparently exactly what happens with a potential wall in graphene and it is called Klein tunneling [2, 3, 5], in analogy with relativistic quantum mechanics. In the present logic, for example in a nanoribbon of graphene where the band gap is not strictly zero at any width, but can come close to it, a potential barrier formed by a gate or substrate could be overcome more easily because of the special topology of the graphene ribbon. But we could also consider a three-dimensional material such as InSb where again the very small gap of 0.14 eV causes, according to Kane (and band theory), a very small effective mass of 0.014 me which makes tunneling through a barrier a very much easier proposition than through a silicon barrier, for example. Basically, it means that the wavepacket attempting to cross has already mixed into the waves which are reappearing on the other side. One can even produce smaller effective masses than InSb in suitably engineered semiconductors containing Hg.

4.3 Graphene properties

Structure

Graphene carbon atoms are densely packed in a hexagonal pattern. Each atom has three sigma-bonds and one pi-bond out of plane, with an interatomic distance of about 1.42 Å. The sp^2 hybridization combined to the tightly packed carbon atoms explain graphene's stability. One of the most extraordinary properties of this material is the ability to self-repair holes in its sheets by reacting with hydrocarbons.

Transport properties

The temperature dependence of graphene mobility (see figure 4.8(a)) is mainly due to acoustic phonon scattering [5, 11]. The mobility can reach values of order 10^5 cm^2 V^{-1} s^{-1} at $T = 300$ K, unique for any material. In two-dimensional GaAs systems, the acoustic phonon scattering is important below $T = 100$ K, and polar optical phonon scattering becomes exponentially more important for $T > 100$ K. In graphene, on the other hand, the linear dependence of T on resistivity is observed up to very high temperature ($T \sim 1000$ K) because the optical phonons which strongly scatter have very high energy ($T \sim 2000$ K) and are not relevant for carrier transport. These findings constitute perhaps the most important features of relevance to fast modern electronics, emphasizing how important it is to get rid of all forms of impurity defect and substrate disorder scattering.

At low temperature, suspended graphene shows a near-ballistic transport over micron dimensions with a very high mobility reaching 170 000 cm^2 V^{-1} s^{-1} [12]. At higher temperatures, the resistivity increases linearly with the temperature, revealing a phonon scattering behavior. The mobility reaches 120 000 cm^2 V^{-1} s^{-1} at 240 K, the highest mobility for a semiconductor.

In two-dimensional electron gases with high mobility, it is common to observe a quantum Hall effect. It has also been observed in graphene; see figure 4.8(b).

The discrete steps in the Hall current are a manifestation of the quantum Hall effect. The Hall current can only flow when the density of states at the Fermi level

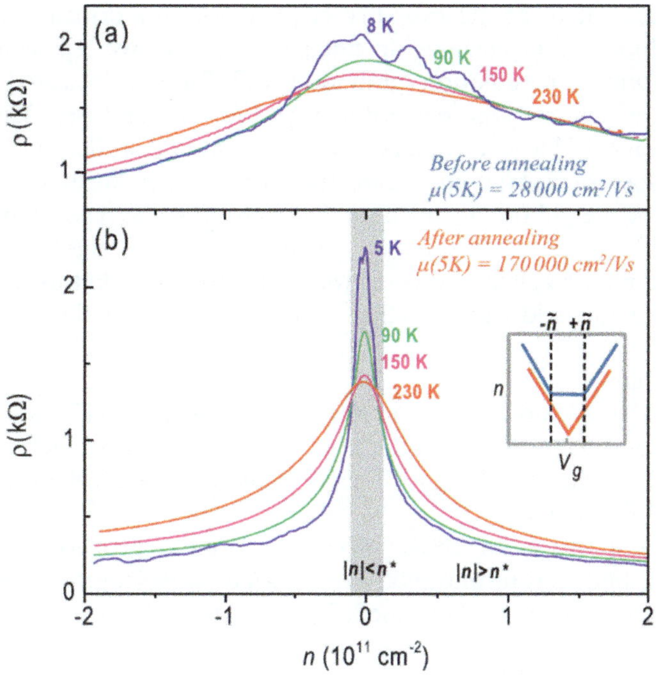

Figure 4.8(a). Temperature dependence of resistance of suspended device before and after current annealing. Inset: sketch of gate voltage dependence of the carrier density in clean and charge inhomogeneous graphene. Reprinted with permission from [11]. Copyright 2008 by the American Physical Society.

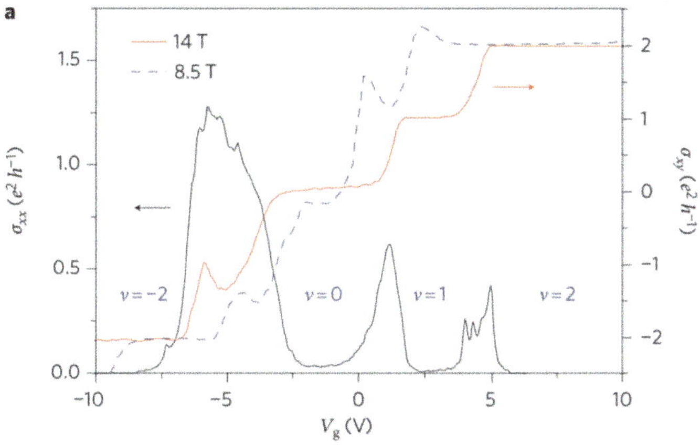

Figure 4.8(b). Longitudinal (left axis) and Hall conductivity (right axis) versus gate voltage at B = 14 T (solid line) and 8.5 T (dashed line), for monolayer graphene. Reprinted [13] Springer Nature, copyright 2010.

reaches a critical value, at the mobility edge. Below the critical value (typically 10^{11} cm^2), the localized eigenstates do not have enough overlap with at least two neighbors to form a percolation path and will form a localized spectrum of eigenstates [14] which carry no current at very low temperatures. The localization is a result of the inevitable, even if very small, amount of disorder in the material. As the density of localized states increases, there comes a point at which the energy difference to the neighboring touching localized state is less than the natural diffusional width of the eigenstate (in two dimensions all eigenstates are in fact localized for an infinite system, however weak the disorder is in this case, it is helped by a strong magnetic field which further reduces the effective dimensionality).

Optical properties
Given the fact that we now have an electrical conductor which can be made as a suspended monolayer, for photoconduction, the next question is how does this layer (s) transmit reflect and absorb light? This is studied in the excellent review by L A Falkovsky [12] (figure 4.9).

Nanotubes and graphene show promising optical properties. Graphene in particular has a higher transmittance over a wider wavelength range than single-walled carbon nanotube films, thin metallic films and indium tin oxide (ITO) [15].

Graphene thermal conductivity
Graphene is in the category of materials with the best thermal conductivity [15–17]. Pristine graphene has values just below diamond. As it turns out, for suspended graphene, both the lattice and the electronic contributions are very high; see figure 4.10.

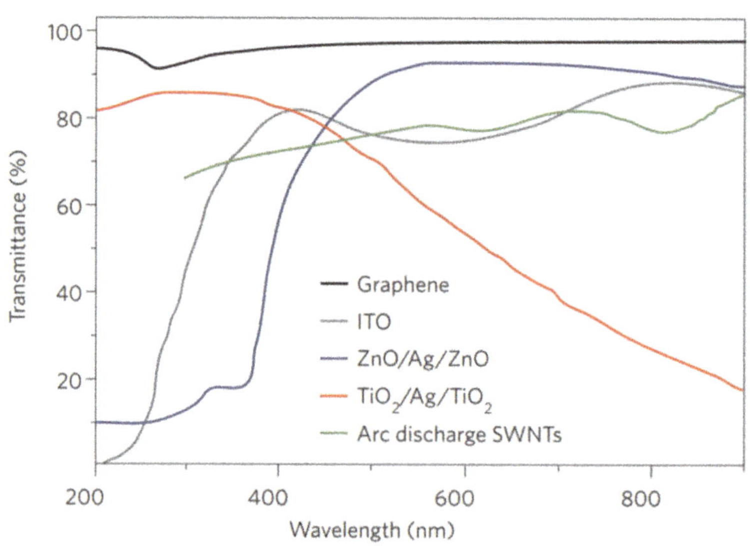

Figure 4.9. Transmittance versus wavelength of graphene, metallic films, indium tin oxide (ITO) and single-walled nanotubes (SWNT). Reprinted from [15] by permission from Springer Nature, copyright 2010.

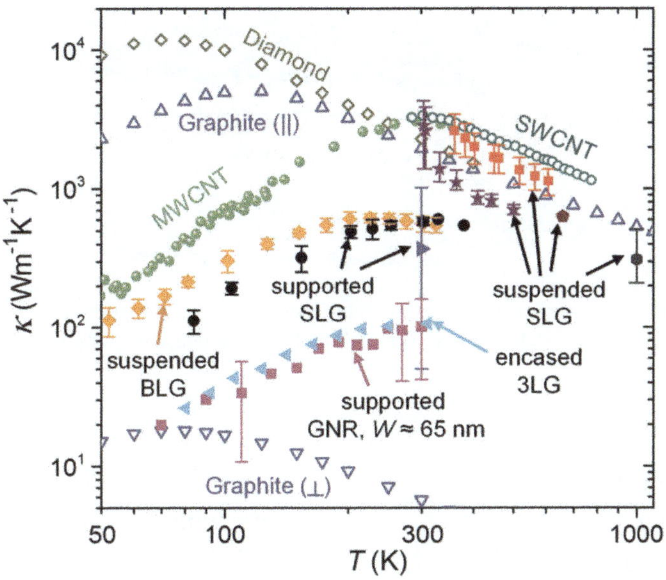

Figure 4.10. Experimental thermal conductivity κ as a function of temperature T. Reprinted from [17] with permission from John Wiley and Sons.

Suspended single-layer graphene (SLG) has values in the range 2000–4000 W mK^{-1} at $T = 300$ K and 700–800 W K^{-1} m^{-1} at $T = 500$ K. SWNT means single-wall nanotube, M stands for multi, GNR is graphene nanoribbon.

The purely electronic contribution has been computed by Tae Yun Kim *et al* and turns out to be as high as that of a metal and about a third of the total thermal conductivity [18]. The diamond-like thermal conduction opens the door to many applications but unfortunately not thermoelectric ones.

Graphene is turning out to be an excellent water filter for water desalination [19].

4.4 Graphite

4.4.1 Structure

Graphite's structure is composed of sheets of graphene that are van der Waals bonded on top of each other (figure 4.11). These bondings are rather weak compared to the covalent bonding in the graphene sheets, which make it easy to exfoliate graphene sheets from graphite. The carbon atoms in each layer are arranged in a honeycomb lattice, the separation of which is 0.142 nm and the distance between planes being 0.335 nm. There are two known forms of graphite, alpha (hexagonal) and beta (rhombohedral). Despite their similar physical properties, the difference comes from the graphene layers which stack slightly differently. The alpha form can be converted to the beta form through mechanical treatment and the beta form reverts to the alpha form when it is heated above 1300 °C.

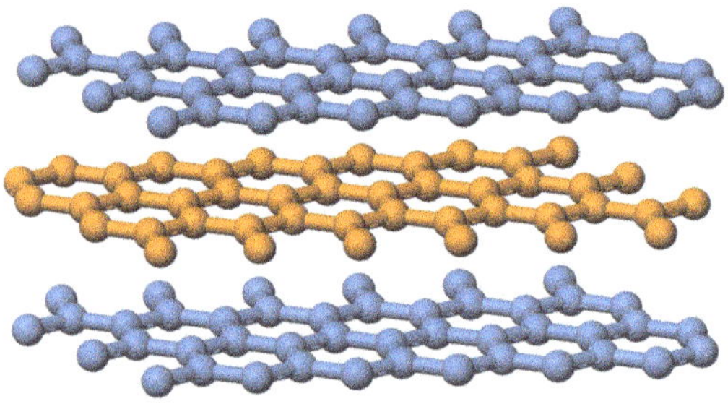

Figure 4.11. Graphite and intercalated graphite. This image has been obtained by the author from the Wikimedia website, where it is stated to have been released into the public domain. It is included within this book on that basis.

4.4.2 Properties and applications

The acoustic and thermal properties of graphite are highly anisotropic. Phonons can propagate quickly along the tightly bound planes, but they are slower when they have to cross from one plane to another.

Graphite is an established electric conductor. It is useful in such applications as arc lamp electrodes. It can conduct electricity because the electrons are delocalized within the carbon layers (a phenomenon called aromaticity). The valence electrons are free to move, and they respond to an applied electric field. Consequently they are able to conduct electricity. The transported charge or electricity is primarily conducted within the plane of the layers. The conductive properties of powdered graphite allow it to be used as a pressure sensor, in particular in carbon microphones.

Graphite is thus a very useful material. It has a wide range of applications in industry, for making batteries, pencils, and electrodes.

4.4.3 Pyrolytic graphite: diamagnetism [2, 3]

Pyrolytic graphite is a synthetic material that is similar to graphite (figure 4.12). It is produced by the decomposition of a hydrocarbon gas at very high temperatures in a vacuum furnace. This process of decomposition permits the graphite to crystallize into a layer-by-layer composition. Pyrolytic graphite has a single cleavage plane similar to mica. Now comes a seriously interesting fact: pyrolytic graphite exhibits the greatest diamagnetism of any solid at room temperature (by weight). The strong diamagnetism makes it possible to levitate thin slices over rare earth magnets. Making strongly diamagnetic materials is one of the targets of materials research for reasons which are fairly obvious since levitation brings with it a whole area of novel applications, in particular frictionless motion (trains, hovercraft, etc). Superconductors are excellent diamagnets, but again here it is the restriction to low temperatures which is a problem. Thus the search for high-T_c superconductors.

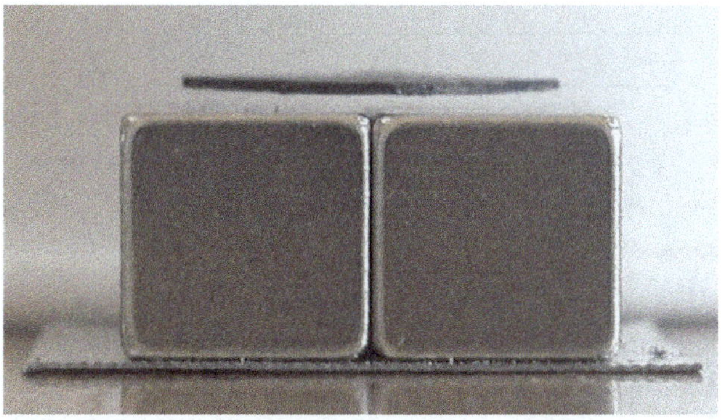

Figure 4.12. Pyrolytic graphite floating on two magnets. Courtesy of Simon Quellen Field.

4.5 Graphene nanoribbons

Graphene's properties can be exploited through graphene nanoribbons (GNR). Based on how these ribbons are cut, there exists two types of GNR, as depicted in figure 4.13.

Note that the armchair structure can go down to zero gap at $k = 0$ for $N = 5$, despite the lateral confinement (figure 4.14). This means that the special features associated with the linear dispersion relation such as ease of tunneling are reachable here in a simpler configuration than graphene, where zero mass behavior is seen near K and K′. One can show using k-p theory that the anomalous low mass, Dirac analogy is a direct result of the low (down to vanishing) band gap. The III–V material InSb has a $k = 0$ gap of 0.14 eV and an effective mass of 0.023 m0. Further, when the gaps are low and indeed lower than spin–orbit energies, some very novel material and spin properties are to be expected. Figure 4.15 shows that the band gap of GNR can be controlled by the width, as predicted theoretically, and can reach very low values down to a few meV.

Graphene transistors

Nanoribbons' shape allow one to exploit graphene properties in usual devices, such as a field effect transistor (figure 4.16).

Large-area graphene transistors have a unique current–voltage transfer characteristic (figure 4.17). The carrier density and the type of carrier (electrons or holes) in the channel are controlled by the potential differences between the channel and the gate. Large positive gate voltages promote an electron accumulation in the channel (n-type channel), and large negative gate voltages lead to a p-type channel. This behavior gives rise to the two branches of the transfer characteristics separated by the Dirac point. Its position depends on several factors: the difference between the work functions of the gate and the graphene, the type and density of the charges at the interfaces at the top and bottom of the channel, and any doping of the graphene. Because of its high carrier mobility, graphene transistors are of particular interest for ultra-high-speed radio-frequency electronics [23].

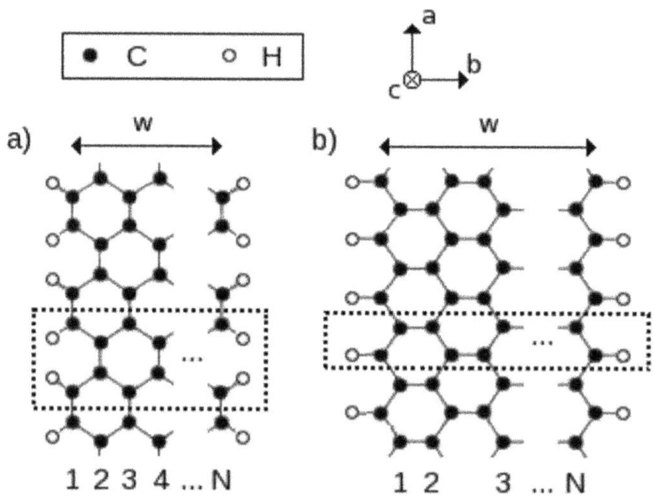

Figure 4.13. Structure of graphene nanoribbons with (a) armchair edges (armchair ribbon) and (b) zigzag edges (zigzag ribbon). A primitive unit cell is marked out in each case, and the ribbons are periodic in the a direction. The number N denotes the number of C atoms in the b direction, and the ribbon width W is given by the maximal distance between C atoms in the b direction. Reprinted with permission from the American Chemical Society [20], copyright 2010 American Chemical Society.

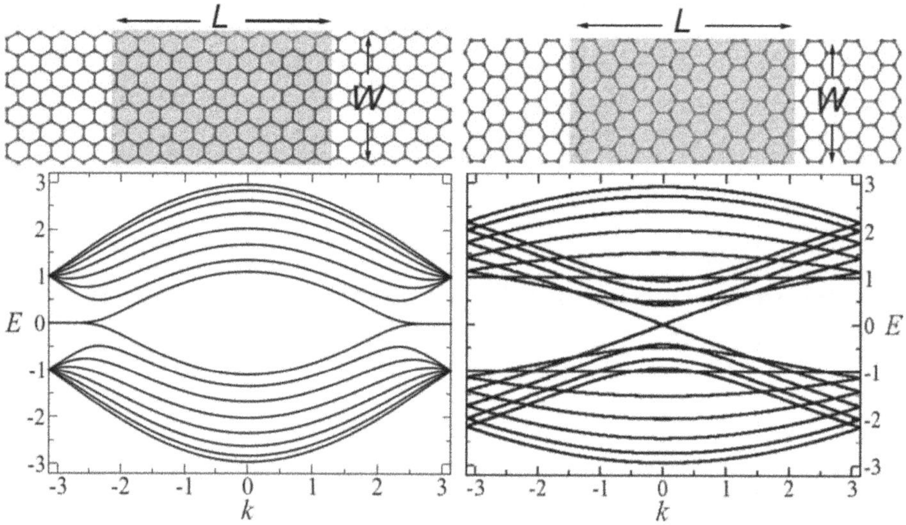

Figure 4.14. Top: graphene nanoribbons with zigzag (left) and armchair (right) edges with disordered regions (shaded areas) of length L and width W attached to perfect leads. Bottom: band structures calculated for the clean zigzag (left) and armchair (right) graphene nanoribbons. Reprinted with permission from the American Chemical Society [21], copyright American Chemical Society.

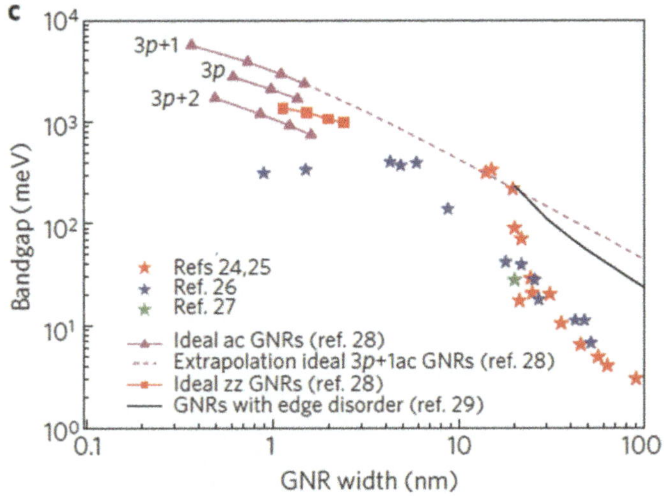

Figure 4.15. Band gap versus nanoribbon width GNR. Reprinted from [38] by permission from Springer Nature, copyright 2010.

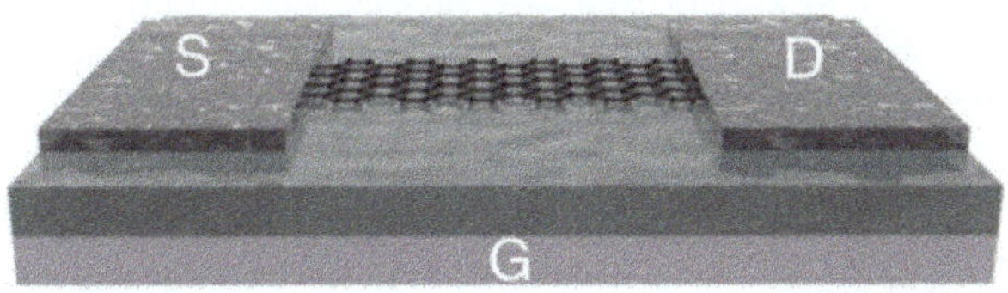

Figure 4.16. Typical nanoribbon device configuration with a short-channel graphene nanoribbon field effect transistor GNRFET with nine atoms GNR channel and Pd source–drain electrodes. From [22] Creative Commons Attribution 4.0 International License.

Graphene as a photodetector

The unique two-dimensional topology and high mobility of graphene even at room temperature make this material very important for photodetector and sensor applications.

Figure 4.18 illustrates a particularly useful and versatile example of a quantum dot (QD) on two-dimensional layer structure where the exciton in the QD is split by hole transfer to the high mobility graphene, with ultra-high gain.

4.6 Graphene oxide fluorescence

Pure pristine graphene does not fluoresce well at all, but when functionalized to form graphene oxide (GO; see figure 4.19), the fluorescence is considerably enhanced. The origin and optimization of the photoluminescence (PL) in GO and related materials is a fascinating topic and not completely understood. As usual with PL, it is expected to give rise to many applications in bio-imaging and monitoring catalysis.

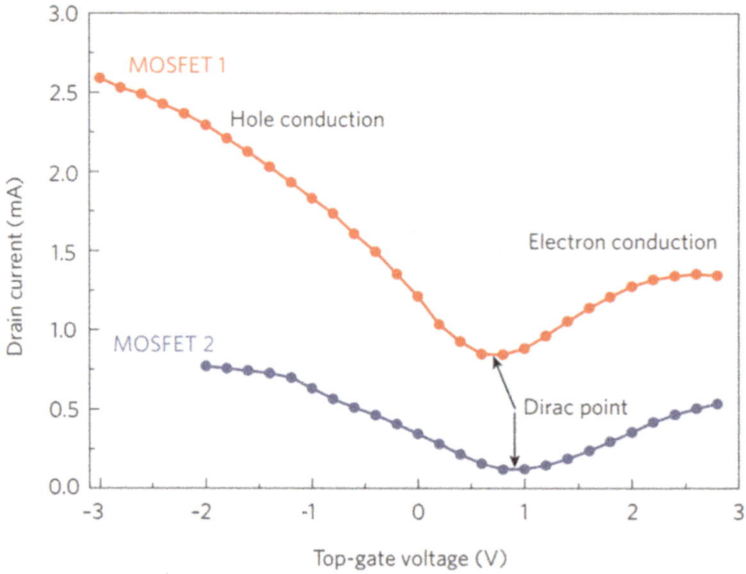

Figure 4.17. Direct-current behaviour of graphene MOSFeTs with a large-area-graphene channel. Typical transfer characteristics for two MOSFETs with large-area-graphene channels. Unlike conventional Si MOSFETs, current flows for both positive and negative top-gate voltages. Reprinted from [38] by permission from Springer Nature, copyright 2014.

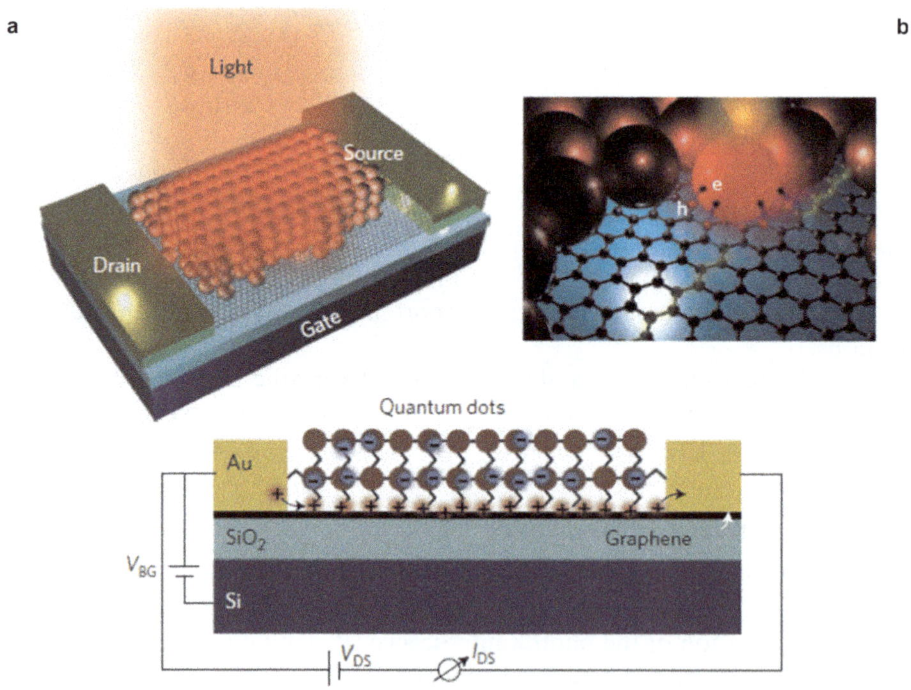

Figure 4.18. Photodetector based on graphene/quantum dots (QDs). Reprinted from [38] by permission from Springer Nature, copyright 2014.

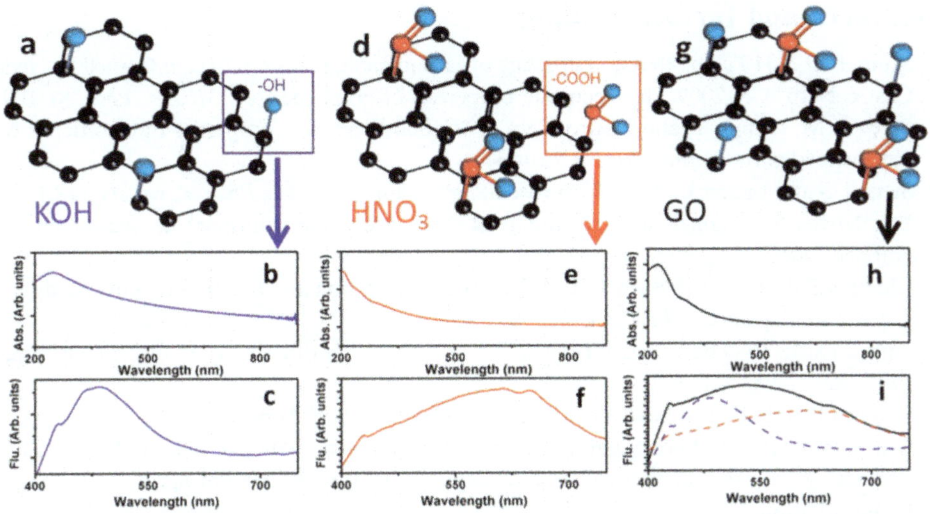

Figure 4.19. Optical properties of functionalized graphene oxide (GO). (a) The OH-rich GO and (d) the COOH-rich GO. (g) GO; the UV visible absorption spectra of (b) the OH-rich GO and (e) the COOH-rich and (h) GO. The fluorescence spectra of (c) the OH-rich GO and (f) the COOH-rich GO and (i) GO. Reprinted with permission from the American Chemical Society [24]. Copyright © 2013 American Chemical Society.

4.7 Summary

Graphene was discovered nearly accidentally, and since then has occupied the mind of a large number of nanotechnology physicists worldwide and has led to a recent Nobel Prize. It is the first high-quality, atomically two-dimensional, band-tunable metal/semiconductor system with the world's highest mobilities at room temperature. This chapter reviewed the basic properties and explained why they come about. The material covers band structure, a few applications and also reviews graphene nanoribbons and bilayers. Material fabrication has become such a powerful and versatile tool, we may ask: what can we learn from graphene-like band structures? Is it worth trying to reproduce them in other materials? What more can we learn about many-body effects? This is perhaps one of the most important questions. This question has now been partially answered by the discovery of Mott–Anderson supeconductivity in twisted graphene bilayers [25] (see chapter 7). Graphene has now been complemented by finite-gap two-dimensional materials made with transiton metal dichalcogenides (TMDC), treated in chapter 7.

4.8 Exercises

(Q1) Using Bloch's theorem solve equations (4.2) and (4.3) and find the dispersion relation ($E(\mathbf{k})$), band structure of graphene.

(Q2) Find the points in k space at which the gap vanishes. Make a rough plot of $E(\mathbf{k})$ and point out where the Fermi energy is.

(Q3) What is special about a semiconductor with very small or vanishing band gap?

References and further reading

[1] Minot E D 2004 Tuning the band structure of carbon naonotubes *PhD Thesis* Cornell University

[2] Castro Neto A H 2009 The electronic properties of graphene *Rev. Modern Phys.* **81** 109

[3] Peres N M, Guinea F and Castro Neto A H 2006 Electronic properties of disordered two-dimensional carbon *Phys. Rev.* B **73** 125411

[4] Stenull O and Lubensky T 2014 Penrose tilings as jammed solids *Phys. Rev. Lett.* **113** 158301

[5] Das Sarma S, Adam S and Hwang E H 2011 Electronic transport in two-dimensional graphene *Rev. Mod. Phys.* **83** 407

[6] Chadrasekahr C M and Bush T 2014 Quantum percolation and transition point of a directed discrete time quantum walk *Sci. Rep.* **4** 6583

[7] Suprio Datta 1980 Simple theory for conductivity and Hall mobility in disordered systems *Phys. Rev. Lett.* **44** 828

[8] Wallace P R 1947 The band theory of graphite *Phys. Rev.* **71** 622

[9] Son Y W *et al* 2006 Energy gaps in graphene nanoribbons *Phys. Rev. Lett.* **97** 216803

[10] Razeghi M *Fundamentals of Solid State Engineering* 3rd edn (Berlin: Springer) page 130 – Dirac equation

[11] Bolotin K, Sikes K, Stormer H L and Kim P 2008 Temperature dependent transport in suspended graphene *Phys. Rev. Lett.* **101** 096802

[12] Falkovsky L 2008 Optical properties of graphene and IV-VI semiconductors *Phys. Uspekhi* **51** 887

[13] Dean C R *et al* 2010 Boron nitride substrates for high quality graphene electronics *Nat Nanotehcnol* **5** 722–26

[14] Movaghar B 1992 *Phil. Mag.* B **65**

[15] Bonaccorso F, Sun Z, Hasan T and Ferrari A 2010 Graphene photonics and optoelectronics *Nat. Photon.* **4** 611

[16] Bonnacorso F *et al* 2015 Review : Graphene related two-dimensional crystal and hybrid systems for energy conversion and storage *Science* **347** 1246501

[17] Xu Y *et al* 2014 Thermal and thermoelectric properties of graphene *Small* **10** 2182–99
See also; Balandin A A 2011 Thermal properties of graphene and nanostructured carbon materials *Nat. Mater.* **10** 569–89

[18] Kim T Y *et al* 2016 The electronic thermal conductivity of graphene *Nanoletters* **16** 2439

[19] Surwade S P *et al* 2015 Water desalination using nanoporous single layer graphene *Nat. Nanotech* **10** 459

[20] Tan Z W *et al* 2011 First-principles study of heat transport properties of graphene nanoribbons *Nano Lett.* **111** 214–9

[21] Kleftogiannis I *et al* 2013 *Conductance through disordered graphene nanoribbons: standard and anomalous electron localization* Phys. Rev. B **88** 205414

[22] Llinas J P *et al* 2013 Shortchannel field-effect transistors with 9-atom and 13-atom wide graphene nanoribbons *Nat. Commun.* **8** 633

[23] Liao L, Lin Y-C, Bao M, Cheng R, Bai J, Liu Y, Qu Y, Wang K L, Huang Y and Duan X 2010 High-speed graphene transistors with a self-aligned nanowire gate *Nature* **467** 305–8

[24] Cushing S K, Li M, Huang F and Wu N 2014 Origin of strong excitation wavelength dependent fluorescence of graphene oxide *ACS Nano* **8** 1002–13

[25] Cao Y *et al* 2018 Unconventional superconductivity in magic-angle graphene superlattices *Nature* **556** 43

See also the recent work; Nam Y *et al* 2018 A family of finite temperature electronic phase transtions in graphene multilayers *Science* **362** 324
[26] Geim A K 2011 Nobel lecture: Random walk to graphene *Rev. Mod. Phys.* **83** 851
[27] Geim and Novoselov 2007 The rise of graphene *Nat. Mater.* **6** 183–91
[28] Cao Y *et al* 2018 Correlated insulator behaviour at half-filling in magic-angle graphene superlattices *Nature* **556** 80
[29] Powell J and Crasemann B 1965 *Quantum Mechanics* (Addison Wesley)
[30] Anderson P W 1958 PRB Absence of diffusion in certain random lattices *Phys. Rev.* **109** 1492–505
[31] Chen S *et al* 2010 Adsorption/desorption and electrically controlled flipping of ammonia molecules on graphene *New J. Phys.* **12** 125011
[32] Wakabayashi K, Sasaki K, Nakanishi T and Enoki T 2010 Electronic states of graphene nanoribbons and analytical solutions *Sci. Tech. Adv. Mater.* **11** 054504
[33] Avouris P 2010 Graphene electronics and photonics properties and devices *Nano Lett.* **10** 4285
[34] De Volder M 2013 Carbon nanotubes: present and future commercial applications *Science* **339** 535
Nygard J 2008 Old nanotubes, new tricks *Nat. Phys.* **4** 266
[35] Avouris P 2010 Graphene: electronic and photonic properties and devices *Nano Lett.* **10** 4285
[36] Chuang S L 1995 *Physics of Optoelectronic Devices* (New York: Wiley)
[37] Schwierz F 2010 Graphene transistors *Nat. Nanotechnol.* **5** 487–96
[38] Koppens F H L *et al* 2014 Photodetectors based on graphene, other two-dimensional materials and hybrid systems *Nat. Nanotechnol.* **9** 780–93

IOP Publishing

The Mystery of Carbon
An introduction to carbon materials
Manijeh Razeghi

Chapter 5

Carbon nanotubes and polyacenes

5.1 Introduction

In this chapter we first review the properties of carbon nanotubes (CNTs), a class of carbon structures almost as exciting as graphene and considered mathematically as 'rolled-up' graphene. Again, we look at the band structure, the tuning of band gaps, and some key applications, such as field effect transistors (FETs) and photovoltaic devices (PVs). CNT technology offers unique possibilities and can be built on (i) single CNT units which are suspended, (ii) tubes sitting on a given substrate, or (iii) bundles of nanotubes (NTs). We then turn our attention to the polyacene molecules (molecules made of benzene rings), hydrogen-terminated carbon-based molecules, and the corresponding aromatic molecular crystals. Finally, we briefly visit the exciting world of liquid crystals based on self-assembly of disc-like graphene core molecules. These can act like molecular wires and transport charge in an orderly way in a liquid crystalline phase.

5.2 Graphene and nanotubes

Going back to the previous chapter, the next task is to examine the one-dimensional band structure of an armchair NT using the two-dimensional band structure of graphene [1–3]. The latter must be supplemented with periodic boundary conditions. The armchair NT is obtained by cutting out a slice from the graphene sheet parallel to the x-axis. The slice has a width w which can be expressed as the length of the so-called wrapping vector $\mathbf{w}$ orientated perpendicular to the tube-axis. For an armchair NT, the wrapping vector is of the form $\vec{w} = N(\vec{a}_1 + \vec{a}_2)$ in general $(\mathbf{n_1 a_1} + \mathbf{n_2 a_2})$ where N is an integer, vectors $\mathbf{a}_n$ are shown in figure 5.1. Usually this is also denoted as an (N,N) tube because the wrapping vector is equal to N times $\mathbf{a}_1$ plus N times $\mathbf{a}_2$. Due to the periodic boundary conditions along the y-direction the wavevector component k_y is quantized. Structures and density of states are shown below.

There are three main categories of CNT, as shown on figure 5.2. Based on their formation from graphene, the traditional and easiest way of computing their band

doi:10.1088/2053-2563/ab35d1ch5

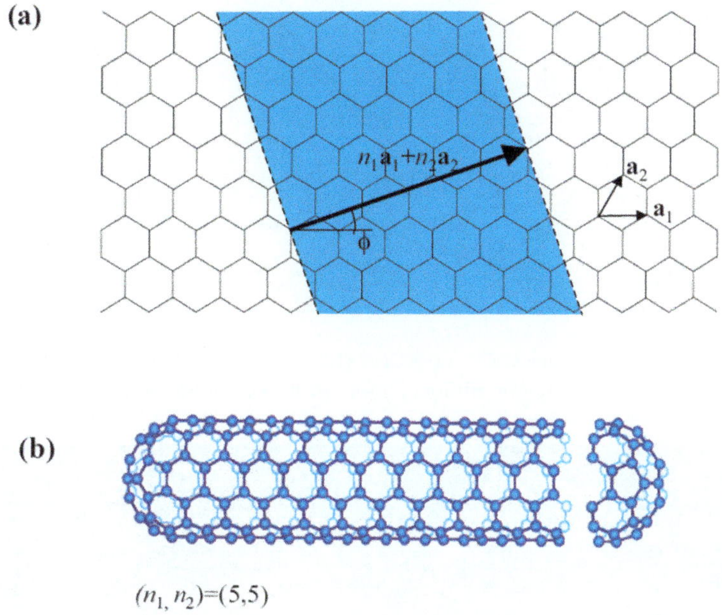

Figure 5.1. Illustrates the formation of a nanotube (NT) from graphene. The graphene lattice and lattice vectors are $\mathbf{a}_1$ and $\mathbf{a}_2$ as shown. A wrapping vector $l = n_1\mathbf{a}_1 + n_2\mathbf{a}_2$ is shown; here it is $l = 4\mathbf{a}_1 + 2\mathbf{a}_2$. The shaded area will be rolled into a tube so that the wrapping vector will encircle the waist of the tube. The chiral angle is measured between the wrapping vector l and $\mathbf{a}_1$. Reprinted with permission from [3].

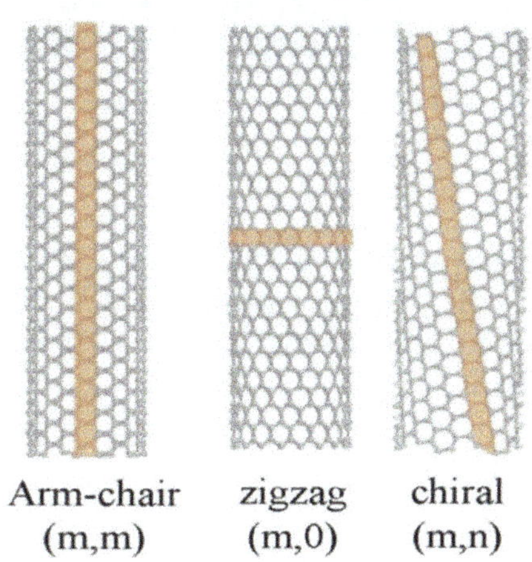

Figure 5.2. Various nanotube (NT) figures showing the most important different types. Reprinted with permission from [4]. Copyright 2014 Elsevier.

structure is to derive it from graphene's (see chapter 4 for more information) (figure 5.3).

One can make single-wall nanotubes (SWNTs) and multi-walled nanotubes (MWNTs), as shown in figure 5.4. The strong covalent bonding gives the CNTs

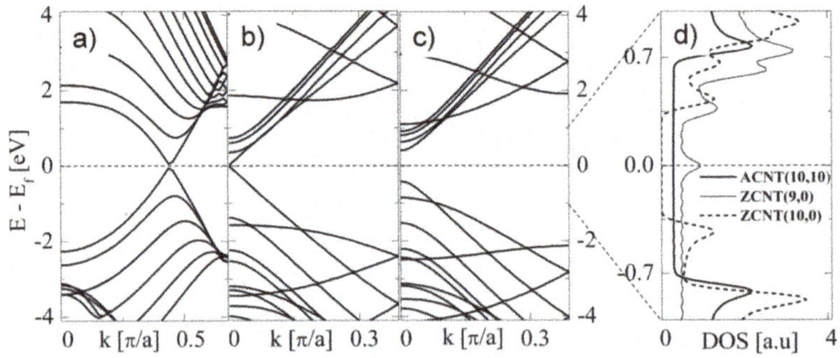

Figure 5.3. Band structure in pristine CNTs. (a) ACNT (10,10), (b) ZNT (9,0), (c) ZNT (10,0), (d) density of states. A = armchair, C = chiral. Reprinted from [2] with permission of Springer Nature, copyright 2011.

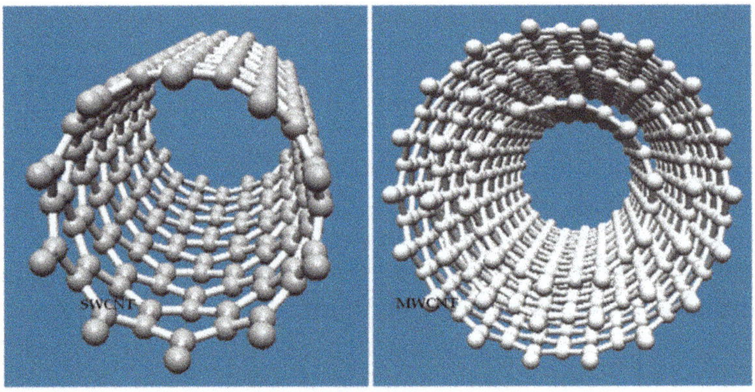

Figure 5.4. Left: Single-walled nanotube (SWNT); right: multi-walled nanotube (MWNT). Reprinted with permission from [5]. Copyright Springer Nature 2019.

high mechanical and thermal stability and resistance to electromigration. Current densities as high as 10^9 A cm^{-2} can be sustained [13].

All chemical bonds of the C atoms are satisfied and there is no need for chemical passivation of dangling bonds as in Si. This implies that in CNT electronics one does not need to use SiO_2 as an insulator. Carrier transport is 1D-like. This implies a reduced phase space for scattering of the carriers and opens the possibility of exploiting ballistic transport. Correspondingly, power dissipation would be lower.

5.3 Carbon nanotube synthesis

The first synthesis of CNTs dates back to 1991, when an electric arc discharge was applied between two graphite electrodes in order to produce carbon fullerenes. There are currently three main methods of synthesis [6].

Electric arc discharge remains popular as it is a rather simple and inexpensive method, and allows one to produce high-quality NTs in quite large quantities. The temperature of the chamber is maintained around 4000 K, while the pressure is usually sub-atmospheric (500 Torr). This method can generate rather high yields, usually 75%, and up to 90% in some experimental conditions. Though MWNTs and SWNTs can be designed this way by using various catalyst precursors, it is difficult to control the chirality of the NTs, which can affect their characterization. Also, another step of purification has to be added to get rid of the metallic catalyst.

Similarly to the electric arc discharge method, laser ablation consists of evaporating graphite with a laser beam and using catalysts to help the synthesis of NTs. The products can be designed in various ways by modifying many experimental parameters, such as the laser power, flow and pressure of the buffer gas, chamber pressure and temperature, or the distance between the target and the substrates. This technique has a high yield and allows the production of low-metallic-impurities NTs. However, the results are usually not uniform, and laser ablation requires high-purity graphite rods and high-power laser beams, and does not produce as many NTs as the electric discharge method, making the latter economically advantageous.

Among all the chemical vapor deposition methods (CVDs), catalytic is the standard technique for the synthesis of CNTs. Metallic nanoparticles are implanted in drills on a silicon substrate, and carbon atoms react with these particles to grow along the drill, at a temperature of approximately 700 °C. This enables one to control the direction of growth and the quality of the NTs. CCVD is therefore an economically practical method for large-scale and high-quality production. The inconvenience of this technique is the purity of the obtained products, which can contain defects.

5.4 Material properties

5.4.1 Thermal conductivity of carbon nanotubes and related materials

CNTs have a remarkably high thermal conductivity which reaches 6000 W mK^{-1} (gold 300 and diamond 2000 W mK^{-1}) and makes them, when combined with good electrical conduction, excellent materials for technology. Much work has therefore been devoted to making CNT polymer composites which combine tensile strength with high thermal conduction and either good or bad electrical conduction. The field is huge, and the reader is referred to the excellent review by Hong *et al* [7].

In some areas such as thermal harvesting or sound insulation, low thermal conductivity is the sought-after target, in which case the composites needed are of the porous type.

5.4.2 Remarkable strength: the space elevator

A space elevator is a theoretical means of space transport. A ribbon-like cable would be attached to the Earth's surface close to the equator and extend to space, beyond the geostationary orbit (around 35 800 km altitude). This design would allow the transport of cargo along the cable, between the Earth's surface and space, without using any rocket launcher. The competition between gravity on the surface and the centrifugal force in space would result in tension of the cable, which would be able to stand up steadily and stationary upon a single position on Earth. Once the so-called

Table 5.1. Comparison of the mechanical properties of CNTs with other strong materials. From Varshney K 2014 Carbon nanotubes: a review on synthesis, properties and applications *International Journal of Engineering Research and General Science* **2**. Reproduced under Creative Commons Attribution 3.0 Unported License.

Material	Young's modulus (Gpa)	Tensile Strength (GPa)	Density (g cm^{-3})
SWNT	1054	150	
MWNT	1200	150	2.6
Steel	208	0.4	7.8
Wood	16	0.008	0.6

tether or ribbon-like cable is deployed, a classic mechanical system would allow climbers to go to space, so that they could deliver cargo to the International Space Station, for example, before simply getting down the same way they went up. This extraordinary proposal does not require any rocket technology, which is remarkable.

Because of CNTs' extraordinary strength, there have been theoretical propositions about the creation of a space elevator. CNTs are many, many times stronger than steel (table 5.1), and buildings built with CNTs would no longer be limited as much in height, since the CNTs are many times stronger than steel, wood, concrete, and other common materials used in today's skyscrapers. In fact, the CNT is so strong, that there is even talk about the potential of a space elevator being built that can welcome passengers into space (figure 5.5).

5.5 Optoelectronic properties

5.5.1 Electrostatic doping and formation of p–n junctions

The need for cheap, easily processed, non-toxic and environmentally friendly technologies has posed CNTs as a promising semiconductor material. In order to exploit its properties, it is necessary to achieve both n- and p-type doping in this material. Doping SWCNTs with acceptor ions is not an issue with the natural adsorption of oxygen by CNT from the environment: the oxygen behaves as an acceptor and is the origin of p-type SWCNT. The real challenge is to dope SWCNT with donor ions and achieve a certain doping stability and an overall performance of the devices [8]. The most effective n-type doping method involves electron transfer reactions, as this elegant technique produces air-stable n-type SWCNTs. Generally, the CNT will be reduced with the electron transfer, resulting in n-type behavior, while the remaining cationic dopants will insulate it from air. These simple methods could easily be implemented into the solar cell manufacturing process through a simple immersion process.

Having both p- and n-type doped material allows the fabrication of basic devices that pave the way to new applications. Ji Ung Lee [9] has thus achieved the fabrication of a p–n junction with SWCNT; see figure 5.6.

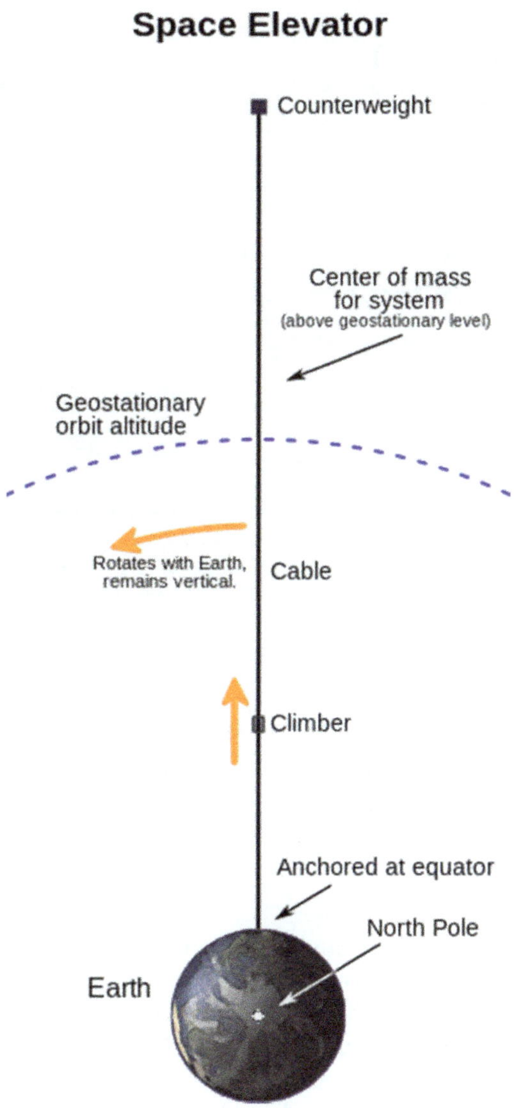

Figure 5.5. Artist impression of a space elevator. This image has been obtained by the author from the Wikimedia website where it was made available by Booyabazooka under a CC BY-SA 3.0 licence. It is included within this book on that basis. It is attributed to Booyabazooka.

5.5.2 Photonic response of a nanotube p–n junction

CNTs have a large number of real and potential speculative applications. The reader is referred to recent review articles and, in particular, [2, 9, 10]. Single NTs can act as photo-responsive elements (see figure 5.7) or as single electron transporters and spin wires. The mobility depends on tube type and purity but can reach high values of several hundred $cm^2\ V\ s^{-1}$, though it must be remembered that for such short lengths of a few microns the concept of mobility does not always apply. Indeed, one can

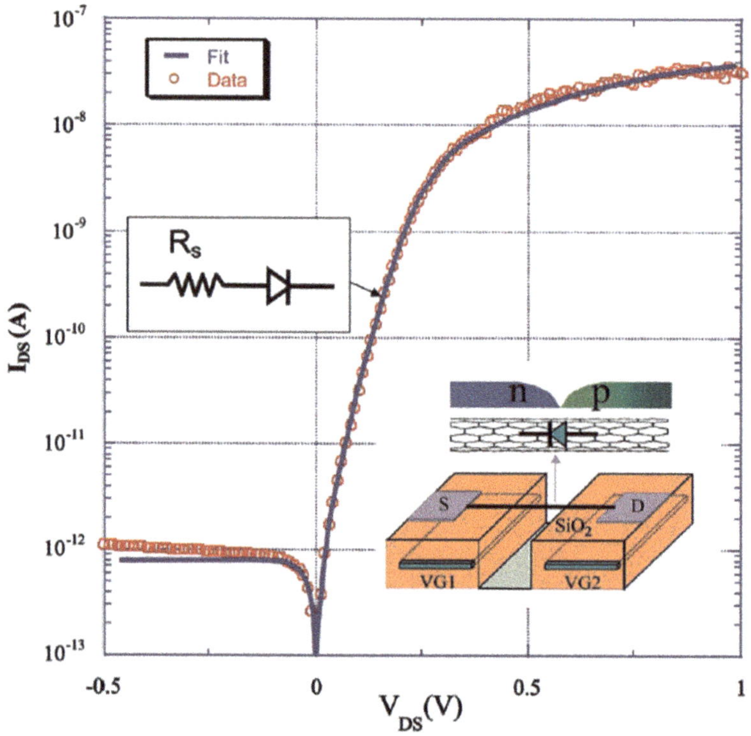

Figure 5.6. The inset shows the split gate device where the gate bias VG1 and VG2 are biased with opposite polarities (VG1 = −VG2 = 10 V) to form an ideal p–n junction diode along a SWNT. Data are typical dark current voltage (*I*–*V*) curve at $T = 300$ K fitted to the p–n model. Reprinted from [9] with the permission of AIP Publishing.

envisage charge carriers to be injected at one end and cross the tube by quantum diffusion while keeping their spin orientation, i.e. quantum diffusion. In these limits the transport is limited by what happens at the contacts and importantly also by electron–electron scattering and spin–orbit scattering. Weak magnetic forces such as spin–orbit coupling, which are normally negligible, can suddenly acquire critical importance and make new science and lead to quantum spintronic applications. CNTs are wonderful objects for studying quantum transport and dissipation. They can be made as metals or as semiconductors; see the band structures in figure 5.3. CNTs can be functionalized in a multitude of different ways for different response applications, in particular to sensors. They now constitute a very important component of materials in nano-engineering. They can be incorporated in building more complex devices including polymers and semiconductors. Here are a few examples, starting with bolometric response (figure 5.8).

This response is in the useful range and can have technical applications. It is also interesting to compare the results obtained with suspended CNTs and substrate-grown CNTs. Mann *et al* have thus compared devices including suspended and substrate-grown SWCNTs (figure 5.9).

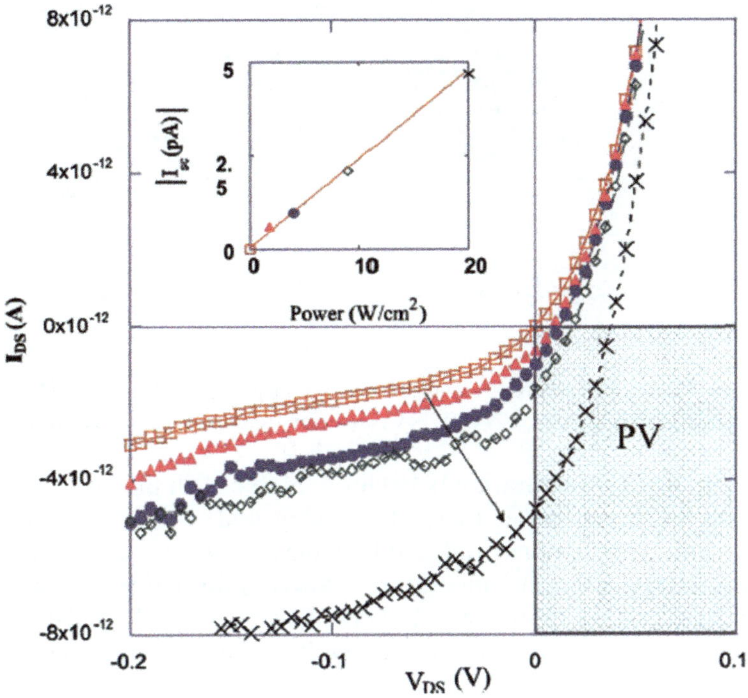

Figure 5.7. I–V characteristic under increased light intensity showing progressive shift into the fourth quadrant (photovoltaic, PV) where the diode generates power. Inset: the expected linear increase in the current measured at $V_{DS} = 0$ (I_{SC}) with illuminated power. Reprinted from [9] with the permission of AIP Publishing.

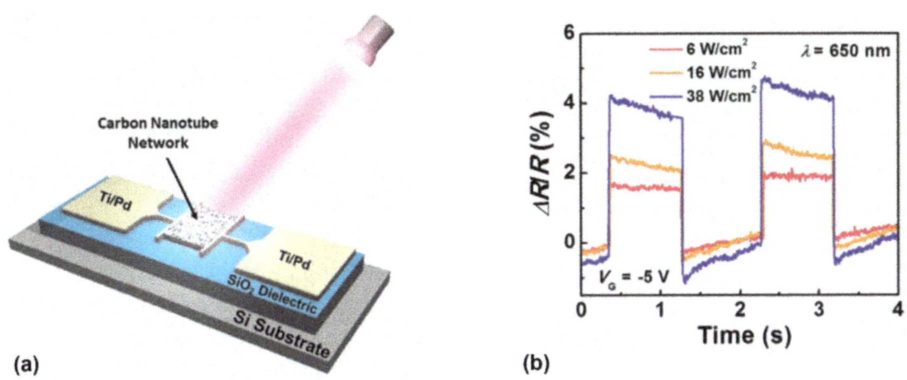

Figure 5.8. Photo-switching characteristic of a single chirality (9,8) nanotube (NT) transistor. Used with permission from [11]. Licensed under a Creative Commons Attribution 4.0 International License.

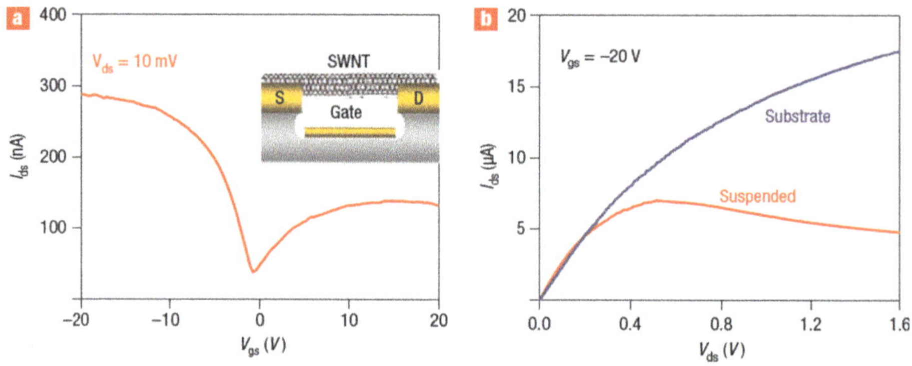

Figure 5.9. Illustrates the current versus gate voltage for a 2 μm-long suspended quasi-metallic-SWNT device (left) versus V_{ds} (right) curve. Reprinted from [57] by permission from Springer Nature, copyright 2019.

Note the very different saturation behavior at high bias in the two cases. The tube on a substrate has much better thermal coupling and a wider three-dimensional phonon spectrum. Hot phonons produced by joule heating are removed very quickly to the substrates. The electron temperature returns to normal very quickly and not many charges are excited to low-mobility channels or traps. This also implies that just touching (thermally contacting) the suspended tube can cause a large change in the current.

5.6 Applications of carbon nanotubes

5.6.1 The carbon nanotube-field effect transistor (CNT-FET) device

CNTs are currently considered as promising building blocks of future nanoelectronic technology. This is not simply due to their small size but rather to their overall properties. In fact, many of the problems that silicon technology is facing, or will face are not present in CNTs. Carrier transport is one dimension, which makes ballistic transport possible. The chemistry of CNTs also requires less processing for devices since there is no need for chemical passivation, and grant devices with strong mechanical and thermal properties, which are much sought after in electronics [13]. Their key dimension is controlled by chemistry which makes them rather easy to build. Phaedon *et al* managed to build a p-type CNFET, as shown in figures 5.10 and 5.11 (from Skakalova *et al*) [13, 14].

The CNT-FET has applications in nanoelectronics as a switch—observe its high on-off ratio—but also as part of sensoric systems in photonic, molecular and biosensing. The bio-compatibility of carbon makes it an excellent technology for life sciences. Temperature increases the source drain current, implying that surface trapping and Schottky barriers are removed by thermal agitation.

CNT technology is also used for memory applications. In CNT-based random access memory (RAM), a CNT mesh is placed between two metallic electrodes. Applying a current then increases the connectivity of the mesh, modifying the overall resistance. This passive device is thus a CNT-based memory element [15]. There is also an interest in CNT for RF amplifiers [16].

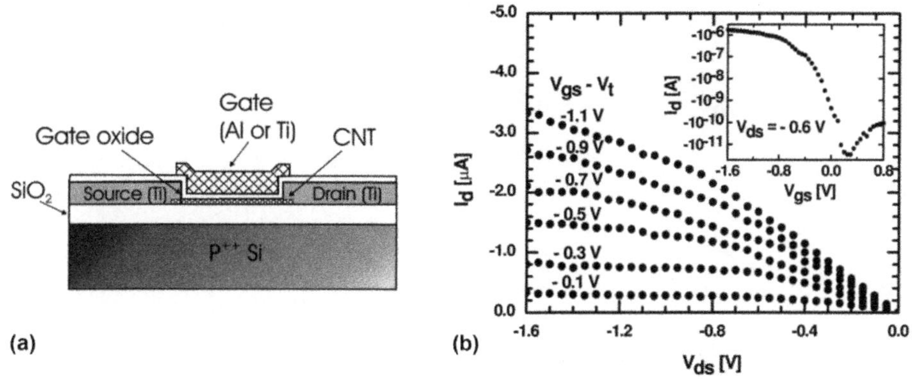

Figure 5.10. Output characteristic of a top gate p-type CNFET with a Ti gate and a gate oxide thickness of 15 nm. The gate voltage values range from 20.1 to 21.1 V above the threshold voltage, which is 20.5 V. Inset: transfer characteristic of the CNFET for V_{ds} = 20.6. Reprinted from [12], with the permission of AIP Publishing.

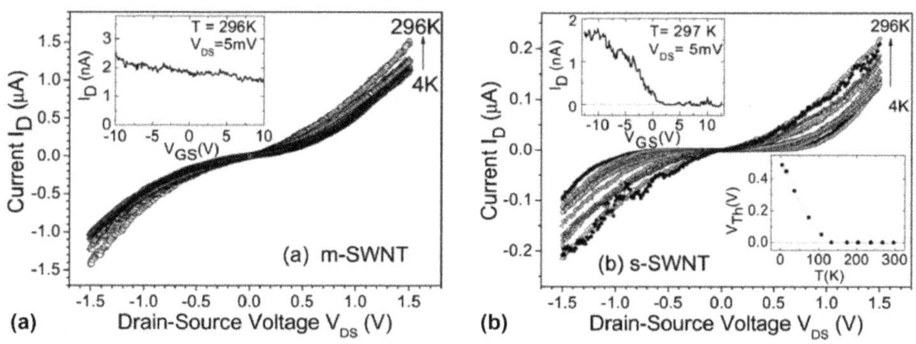

Figure 5.11. Current I_D voltage V_D characteristics of single-walled nanotubes (SWNTs) in a range of temperatures between room temperatures and 4 K. Left curves: metallic SWNT (m-SWNT); right curve: semiconducting SWNT (s-SWNT). Inset: Gate voltage dependence for fixed source drain voltage and temperature. Reprinted with permission from [14]. Copyright 2006 by the American Physical Society.

5.6.2 Carbon-fiber-reinforced polymer technology

Carbon-fiber-reinforced plastic or carbon-fiber-reinforced thermoplastic (CFRP, CRP, CFRTP), often simply called carbon fiber, is a polymer containing carbon fibers and showing an outstanding mechanical resistance while being very light. There are many applications of CFRPs, usually in areas where a high strength-to-weight ratio and rigidity are required, for instance in aerospace, the automotive industry, or in sports equipment. They can, however, be expensive to produce. The most common binding polymer used is often a resin such as epoxy, but sometimes other thermoplastic polymers (polyester, vinyl ester or nylon) are used. The resulting composite may also incorporate various kinds of fibers, such as kevlar, twaron, aluminium, ultra-high-molecular-weight polyethylene (UHMWPE) or glass fibers.

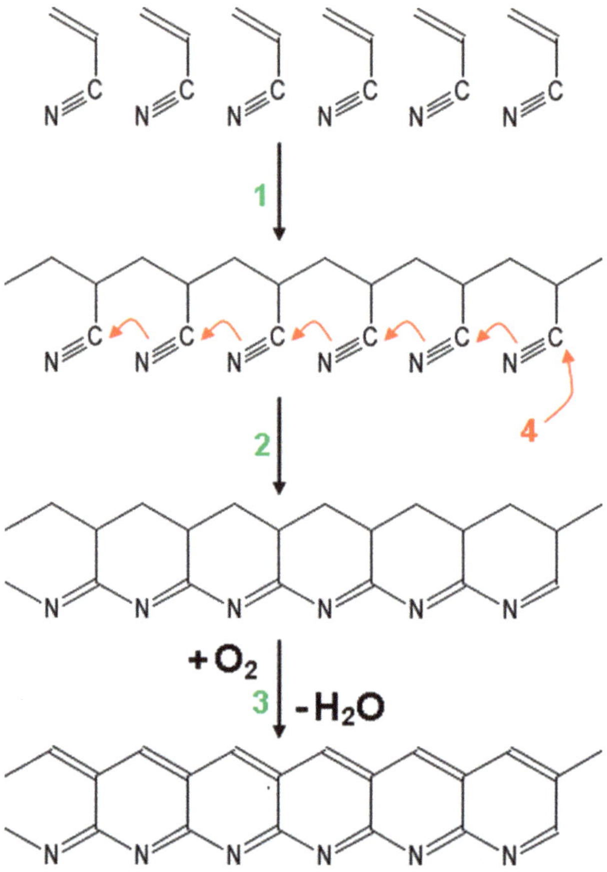

Figure 5.12. Polymerization of acrylonitride through partial oxidation polyacrilonitride (PAN) [17]. This image has been obtained by the author from the Wikimedia website where it was made available by Timur lenk under a CC BY-SA 3.0 licence. It is included within this book on that basis. It is attributed to Timur Lenk.

The most common additive used is silica, but rubber and CNT can also be used. This material is therefore called graphite-reinforced polymer or graphite-fiber-reinforced polymer.

Carbon fibers can be synthesized using polyacrilonitride (PAN); see figure 5.12 below.

Even though they demonstrate incredible and attractive mechanical and chemical properties, carbon fibers are relatively expensive to produce, especially when compared to glass fibers or plastic fibers.

5.6.3 Carbon nanotubes as filters

CNTs have been reported to transport water molecules with orders of magnitude higher efficiency than expected. The experimental results at this time exhibit a wide spread and large-scale computer simulations are being used to determine the most likely values. The NTs could potentially act as filters for water cleansing and desalination [18].

Using CNT membranes as a filter medium is a realistic prospect. It has been shown both by experiment and simulation that water molecules flow through such membranes extremely quickly, which makes them interesting as filters for cost-efficient seawater desalination plants: water molecules pass through the ultra-narrow pores, salt ions do not. This potential for CNT membranes is thus extremely important and being researched intensively.

Water filters have also been recently made using graphene; this is promising to be a very important technology [19].

5.7 Carbon nanodots

Recently researchers have synthesized a new class of nanomaterials called carbon nanodots (CDs) made of pure carbon, which originally came about as a residue in the synthesis of CNTs. At this time no particular structure can be assigned to the CD except that they are three-dimensional, but with mixed bonding and with sizes down to 2 nm. One method of synthesis which may illustrate the outcome intuitively is the preparation by laser ablation of graphite. But there are many other ways, chemical methods and otherwise. The significant novel property which has attracted recent attention to this material is the fact that when surface passivated, the nanodot CD exhibits efficient luminescence (wavelength 400 to 700 nm) which it does not have with an unpassivated surface. This property has given rise to considerable interest, and CDs have become a major modern field of research and development. They combine several favorable attributes of traditional semiconductor-based quantum dots, including size, wavelength dependent luminescent emission, resistance to photobleaching, ease of bioconjugation, without the problems of toxicity or scarcity. They can be produced inexpensively and on a large scale. The photoluminescence quantum yield (QY) is routinely low, around 10%, but various passivation and purification strategies have been found which raises the QY to 50%. The improvement of QY using simple and inexpensive processes is still an issue [20] (figure 5.13).

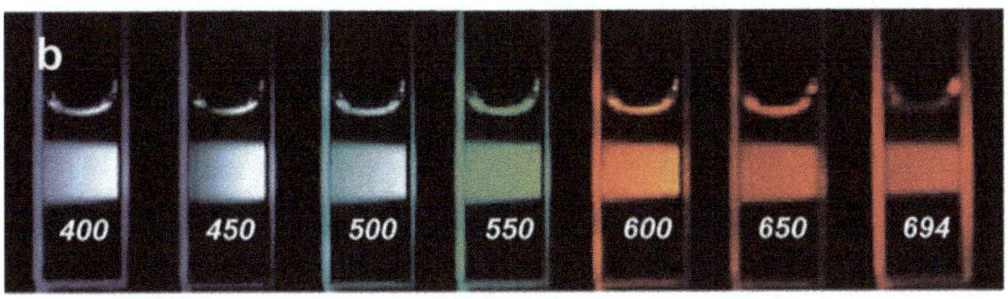

Figure 5.13. Aqueous solution of carbon nanodots (NDs) excited at the indicated wavelengths. Reprinted with permission from [20]. Copyright 2006 American Chemical Society.

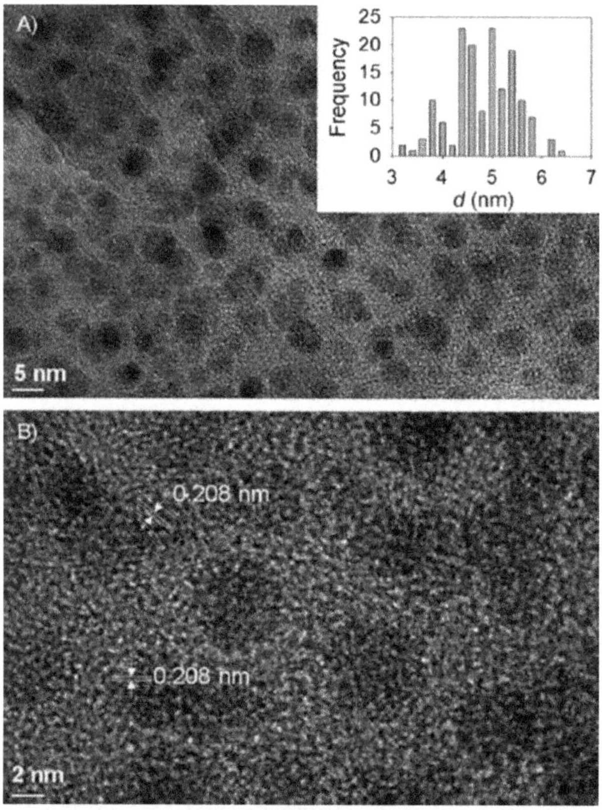

Figure 5.14. TEM micrographs of C-dots at (A) low and (B) high resolution. The inset depicts the particle size histogram. Reprinted with permission from [46]. Copyright © 2009 American Chemical Society.

5.8 Carbon-based molecules with aromatic cores and hydrogen or more complex terminated side chains

In the interesting and vast category of hydrogen-terminated carbon molecules, we find for example benzene (one ring) naphtalene (two rings) anthracene (three rings), tetracene (four rings) and pentacene (five rings) (see figure 5.14), i.e. the pentacenes and higher but less important polyacenes. These molecules can form well-ordered crystals and liquid crystals (with wiggly side chains attached). The crystals form wide band gap semiconductors; the liquid crystals can form semiconductors [49] as columnar (discotic) structures or as layered structures [21]. It is a huge subject, and the reader is referred to the book by Pope and Swenberg [22] for molecules and crystals and specialized literature for liquid crystals [49] (figure 5.15).

The molecular and crystal band gap decreases with ring number, i.e. the longer the molecule, the weaker the confinement and the smaller the band gap (pentacene gap 1.64 eV). The crystals' energy band gaps are controlled by the molecular LUMO and HOMO level separation, because these gaps are normally much bigger than the

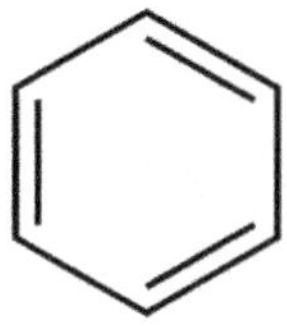

(a) Benzene

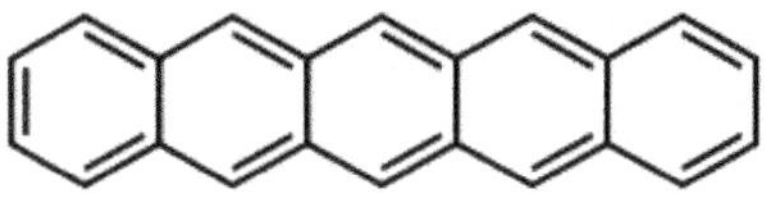

(b) Pentacene

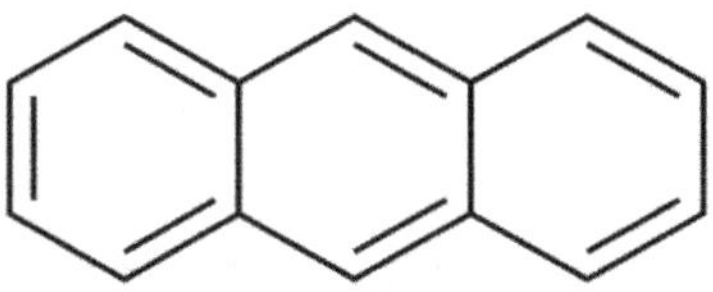

(c) Anthracene

Figure 5.15. Some polyacenes. (a) Benzene, (b) pentacene, (c) anthracene.

bandwidths generated by molecule–molecule coupling (~150 meV). For electronic applications, pentacene has proven to be quite useful: it has a reasonable size gap (1.6 eV), it can be doped, bulk and surface, electrostatically gated, and forms an interesting class of FET [23]. Polaron binding energies can be substantial in these materials: they can reach 200 meV and often exceed the bandwidth in the crystal (150 meV), giving rise to low-mobility polaron hopping transport rather than high-mobility band-like transport (see Coropceanu *et al* [24]). A polaron is a quasi-particle formed when a charge pulls in neighboring ions and thus deforms its environment, digs in, lowers its energy, and in this way creates a new combined entity which we call a polaron. Polarons are a big field in science and the interested reader should consult the monograph by David Emin [25] (figure 5.16).

The high and remarkable low temperature mobility shown in figure 5.17 is a consequence of the ultra-high purity and the suppression of optic phonon scattering by going to low temperatures. This demonstrates that high mobility is not limited to covalently bonded structures such as the III–V compounds, provided the purity is high enough. The temperature sensitivity is due to electron–phonon scattering in a narrow energy band. The optic phonon scattering kicks in as soon as the optic phonon modes start to be excited because of temperature. One way forward would be to engineer molecular crystals which have much higher energy optic modes. This is the case for graphene, for example.

5.9 Molecules with graphene cores and self-assembly

The electronic transport along the columnar stacks made of such disc-like molecules [53] and some others, in particular the triphenylenes (Boden *et al* up to 1 $cm^2\ V\ s^{-1}$ [28]) has been observed and characterized, but planar motion in large core systems

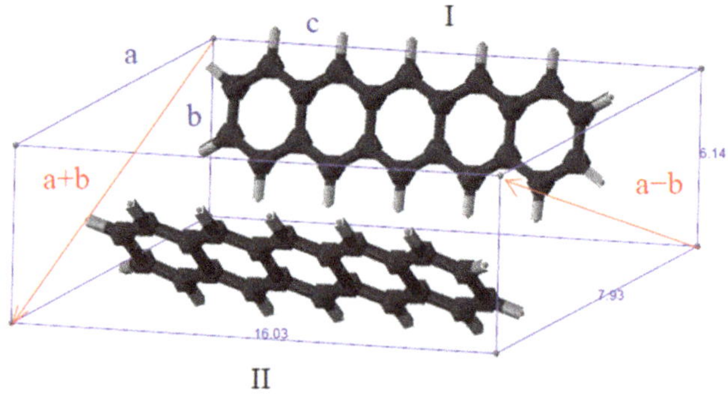

Figure 5.16. Herringbone packing of two pentacene molecules in the unit cell of the Campbell model. The Bravais lattice vectors are **a,b,c**. Reprinted from [26]. Copyright 2004, with permission from Elsevier.

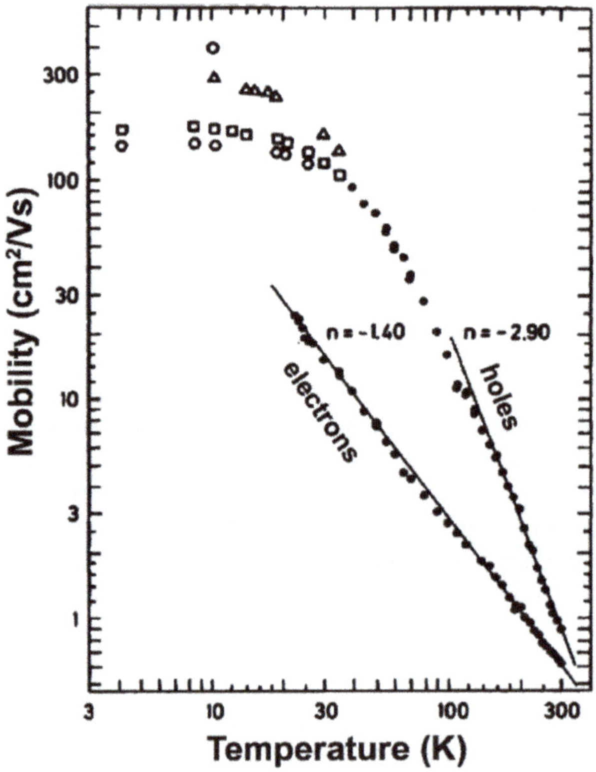

Figure 5.17. Log–log plot of the electron and hole mobilities in ultra-pure naphthalene as a function of temperature. The applied electric field is approximately parallel to the crystallographic **a** direction. Reprinted with permission from [27]. Copyright 1985 by the American Physical Society.

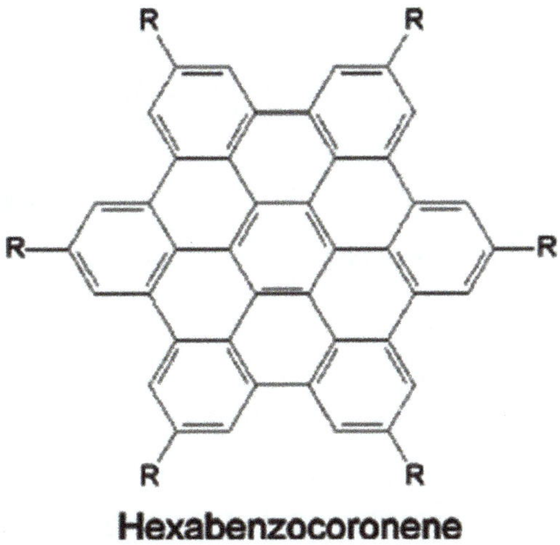

Figure 5.18. Graphene-like core; molecules stack to form columns.

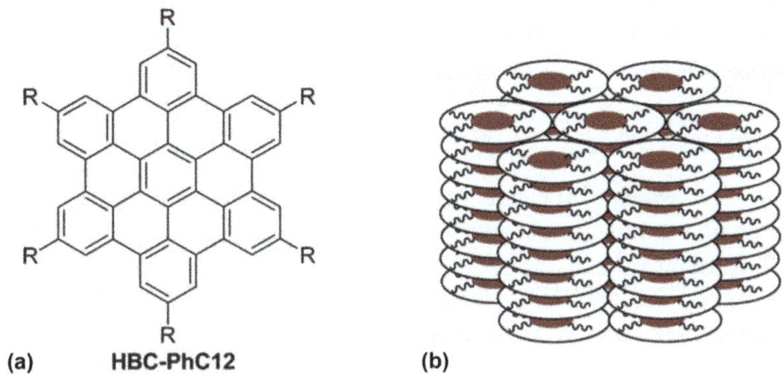

Figure 5.19. Discotic molecules which form self-organized photoconductive columns (right) with liquid crystal PVC efficiencies of up to 2%. Reprinted with permission from [30]. Copyright 2011 American Chemical Society.

has not yet been examined on such a small scale (figure 5.18). It would need nm scale electrode technology, such as scanning tunneling microscope. One can envisage nowadays synthesizing nanowires to replace the side chains and perhaps achieve a new type of controlled self-assembly of addressable devices rather than molecules [29]. Applications of discotic liquid crystals has been discussed by Boden *et al* [29]. The main application of liquid crystals is in display technology, but this cannot be touched upon in this short review (figure 5.19).

A light pulse applied to one electrode generates charge carriers within the penetration depth. The photo charges move in response to an applied field, travel to the counter-electrode. The shoulder in the left picture of figure 5.20 defines the demarcation time at which the charge carrier arrives at the electrode where they are absorbed. This point moves to the left as the field is increased. In the right picture, on

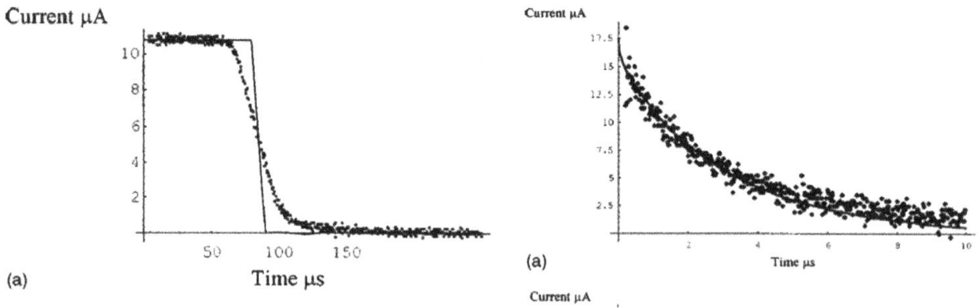

Figure 5.20. Transit signal in the liquid crystalline (80 °C) phase (left) and crystalline phases of HAT6 (20 °C) (right); the applied field is 5 MV m^{-1}. Reprinted with permission from [28]. Copyright 1998 by the American Physical Society.

the other hand, there is no arrival shoulder, meaning that the carriers have a broad distribution of arrival times because they are getting trapped by structural crystalline defects. The important point here is that in the liquid crystalline phase, the structural disorder repairs itself on a short timescale characterized by the motion of the molecules in the discotic columns. So temporarily trapped carriers get back on track quickly enough to give every carrier roughly the 'same' arrival time [31].

5.10 Summary

In this chapter we reviewed the structural properties of carbon nanotubes (CNTs), a class of carbon structures almost as exciting as graphene and considered mathematically as rolled-up graphene. Depending on how the tube is rolled up, it can be metallic or semiconducting. We looked at the different band structures, the tuning of band gaps, and considered some key applications. In particular, we focused on the CNT field effect transistor, and the unique gate voltage and temperature dependences of its characteristics. We then turned our attention to polyacene molecules, hydrogen-terminated carbon-based molecules, and the corresponding molecular crystals. Finally, we visited the exciting world of liquid crystals based on self-assembly of disc-like graphene core molecules. These can act like molecular wires and transport charge in a liquid crystalline phase. In the next chapter, we will turn our attention to the exciting field of conjugated polymers [32, 33].

5.11 Exercises

(Q1) Explain what is meant by fluorescence and discuss its efficiency when compared to nonradiative recombination. Can you think of why pristine graphite does not fluoresce but when chemical complexes are formed fluorescence can be quite strong? If you knew the valence and conduction band wavefunctions, how would you calculate the fluorescence efficiency?

(Q2) The origin of the band gap in organic crystals such as anthracene and naphtalene bonded by van der Waals and dipolar forces is very different from those of covalent structures such as Si and Ge. Explain where the banding and band gaps come from in the organic case. What is a typical bandwidth foranthracene?

(Q3) A charged defect is more efficiently screened in Si than in anthracene. Why? What happens in doped graphene? Remember graphene is 2D-metallic.

References and further reading

[1] Peres N M, Guinea F and Castro Neto A H 2006 Electronic properties of disordered two-dimensional carbon *Phys. Rev.* B **73** 125411

[2] De Volder M F L, Tawfick S H, Baughman R H and Hart A J 2013 Carbon nanotubes: present and future commercial applications *Science* **339** 535–9
Nygard J 2008 Old nanotubes, new tricks, *Nat. Phys.* **4** 266

[3] Minot E D 2004 Tuning the band structure of carbon nanotubes *Thesis* Cornell University

[4] Boumia L, Zidour M, Benzair A and Tounsi A 2014 A Timoshenko beam model for vibration analysis of chiral single-walled carbon nanotubes Physica *E* **59** 180–91

[5] Razeghi R 2019 *Funadamentals of Solid State Engineering* 4th edn (Berlin: Springer)

[6] Eatemadi A, Daraee H, Karimkhanloo H, Kouhi M, Zarghami N, Akbarzadeh A, Abasi M, Hanifehpour Y and Woo Joo S 2014 Carbon nanotubes: properties, synthesis, purification, and medical applications *Nanoscale Res. Lett.* **9** 393

[7] Hong J, Wha Park D and Eun Shim S 2010 A review on thermal conductivity of polymer composites using carbon-based fillers: carbon nanotubes and carbon fibers *Carbon Lett.* **11** 347

[8] Brownlie L and Shapter J 2018 Advances in carbon nanotube n-type doping: Methods, analysis and applications *Carbon* **126** 257–70

[9] Li J U 2005 Photovoltaic effect in ideal carbon nanotubes diodes *Appl. Phys. Lett.* **87** 073101

[10] Baskin A and Kral P Electronic structures of porous nanocarbons *Sci. Rep.* **1** 36

[11] Zhang *et al* 2015 Bolometric-based wavelength-selective photodetectors using sorted single chirality carbon nanotubes *Nat. Sci. Rep* **5** 17883

[12] Wind S J, Appenzeller J, Martel R, Derycke V and Avouris P H 2002 Vertical scaling of carbon nanotube-field effect transistors using top gate electrodes *Appl. Phys. Lett.* **80** 3817

[13] Avouris P *et al* 2003 Carbon nanotube electronics *Proc. IEEE* **91** 1772

[14] Skakalova V *et al* 2006 Electronic transport in carbon nanotubes: From individual nanotubes to thin and thick networks *Phys. Rev.* B **74** 085403

[15] Rueckes T, Kim K, Joselevich E, Tseng G Y, Cheung C and Lieber C M 2000 Carbon nanotube-based nonvolatile random access memory for molecular computing *Science* **07** 94–9

[16] Philbert F and Marsh *et al* 2019 Solving the linearity and power conundrum: carbon nanotube RF amplifiers *Microw. J.* **62** 6

[17] Gupta A K *et al* 1998 Melting behavior of acrylonitride polymers *J. Appl. Sci.* **70** 2703

[18] Shearer C *et al* 2010 Water transport through nanoporous materials porous silicon and single walled carbon nanotubes *Proc. IEEE* **978** p 196

[19] Surwade S P *et al* 2015 Water desalination using nanoporous single-layer graphene *Nat. Nanotechnol.* **10** 459

[20] Sun *et al* 2006 Quantum-sized carbon dots for bright and colorful photoluminescence *J. Am. Chem. Soc.* **128** 7756–7

[21] Funahashi M, Zhang F, Tamaoki N and Hanna J 2008 Ambipolar transport in the smectic E phase of 2-propyl-5′-hexynylterthiophene derivative over a wide temperature range *Chem. Phys. Chem.* **9** 1465–73

[22] Pope M and Swenberg S 1984 Electronic processes in organic crystals and polymers *Ann. Rev. Phys. Chem.* **35** 613–55

Pope M and Swenberg S 1999 *Monographs on the Physics and Chemistry of Materials* 2nd edn (Oxford: Oxford University Press)

[23] Haehlen T, Vanoni C, Waekelin C, Jung T A and Tsujino S 2012 Surface doping in pentacene thin film transistors with few monolayer thick channels *Appl. Phys. Lett.* **101** 033305

[24] Coropceanu V, Cornil J, da Silva D A, Olivier Y, Silbey R and Bredas J L 2007 *Chem. Rev.* **107** 926

[25] Emin D 2012 *Polarons* (Cambridge: Cambridge University Press)

[26] Endres R G, Fong C Y, Yang L H, Witted G and Wöll C H 2004 Structural and electronic properties of pentacene molecule and molecular pentacene solid *Comput. Mater. Sci.* **29** 362–70

[27] Warta K W and Karl N 1985 Hot holes in naphthalene: High, electric-field-dependent mobilities *Phys. Rev.* B **32** 1172

[28] Boden N *et al* 1998 Charge dynamics and recombination dynamics in discotic liquid crystals *Phys. Rev. B* **58** 3063

[29] Boden N, Bushby R, Clements J, Movaghar B and Mater J 1999 Device applications of charge transport in discotic liquid crystals *J. Mater. Chem.* **9** 2081

[30] Kaafarani B R 2011 Discotic liquid crystals for opto-electronic applications *Chem. Mater.* **23** 379–96

[31] Adam D *et al* 1993 Transient photoconductivity in a discotic liquid crystal *Phys. Rev. Lett.* **70** 457

[32] Su W, Schrieffer J R and Heeger A J 1979 Solitons in polyatcetylene *Phys. Rev. Lett.* **42** 1698

[33] Kivelson S and Heim D 1982 Hubbard versus Peierls *Phys. Rev.* B **26** 4278

[34] Hornbaker D J 2000 Electronic structure of carbon nanotubes systems measured with scanning tunneling microscopy *PhD Thesis* University of Illinois at Urbana Champaign

[35] Nygard J 2008 Old nanotubes, new tricks *Nat. Phys.* **4** 266

[36] Movaghar B and Schirmacher W 1981 On the theory of hopping conductivity in disordered systems *J. Phys. C: Solid State Phys.* **14** 859

[37] Wu Y H, Yu T and Shen Z X 2010 Two dimensional carbon nanostructures Fundamental properties synthesis characterization and potential applications *J. Appl. Phys.* **108** 071301

[38] Keller B D, Ferralis N and Grossman J C 2016 Rethinking coal thin films of solution processed natural carbon nanoparticles for electronic devices *Nanoletters* **16** 2951

[39] Lee J U, Gipp P and Heller C M 2004 Carbon nanotube p-n unction *Appl. Phys. Lett.* **85** 145

[40] Hummelen J C, Wudl F and Heeger A J 1999 *Science* **270** 1789

[41] Itkis M *et al* 2006 Infrared photoresponse of suspended single walled carbon nanotubes films *Science* **312** 412

[42] Mann D *et al* 2007 Electrically driven thermal light emission from individual single walled carbon nanotubes *Nat. Nanotechnol.* **2** 33

[43] Walther J H, Ritos K, Cruz-Chu E R, Megaridis C M and Koumoutsakos P 2013 Barriers to superfast water transport in carbon nanotube membranes *Nano Lett.* **13** 1910–4

[44] Li H, Kang Z, Liu Y and Lee S-T 2012 Carbon nanodots: synthesis, properties and applications *J. Mater. Chem.* **22** 24230

[45] Baker S and Baker G A 2010 Luminescent carbon nanodots: emergent nanolights *Ang. Chem. Int. Ed. Nanotechnol.* **49** 6726

[46] Tian L, Ghosh D, Pradhan S, Chang X and Chen S 2009 A review on thermal conductivity of polymer composites using carbon based fillers: carbon nanotubes and carbon fibers *Chem. Mater.* **21** 2803

[47] Keller B D, Ferralis N and Grossman J C 2016 Rethinking coal: thin films of solution processed natural carbon nanoparticles for electronic devices *Nanoletters* **16** 2951

[48] Cushing S, Li M, Huang F and Wu N 2014 Origin of strong excitation wavelength dependent fluorescence of graphene oxide *ACS Nano* **8** 1002

[49] Bushby R, Kelly S and O'Neill M (ed) *Liquid Crystalline Semiconductors* (Springer Series in Material Science vol 169) (Dordrecht: Springer)

[50] Iino H and Hanna J 2005 Electronic and ionic transports for negative charge carriers in smectic liquid crystalline photoconductor *J. Phys. Chem.* B **109** 22120–5

[51] Iino H, Hanna J, Bushby R, Movaghar B, Whitaker J and Cook m 2005 Very High time of flight mobility in the columnar phase of a discotic liquid crystal *Appl. Phys. Lett.* **87** 132102

[52] Gross L, Mohn F, Moll N, Schuler B, Criado A, Guitian E, Pena D, Gourdon A and Meyer G 2012 Bond-order discrimination by atomic force microscopy *Science* **337** 1326

[53] Schmidt Mende L, Fechtenkoetter A, Muellen K, moons E, Friend R H and Mackenzie J D 2001 Selforganised discotic liquid crystals for high efficiency organic Photovoltaics *Science* **293** 1191

[54] Pecchia A *et al* 2002 Photoconductive transient and one dimensional charge carrier dynamics in discotic liquid crystals *Phys. Rev. B* **65** 104204

[55] Crawley D, Nikolic K and Forshaw M (ed) 2005 *3D Nanoelectronic Computer Architecture and Implementation* (Boca Raton, FL: CRC Press)

[56] Endres R G, Fong C Y, Yang L H, Witte G and Wöll C 2004 Structural and electronic properties of pentacene molecule and molecular pentacene solid *Comput. Mater. Sci.* **29** 362

[57] Mann D *et al* 2007 Electrically driven thermal light emission from individual single-walled carbon nanotubes *Nat. Nanotechnol.* **2** 33

IOP Publishing

The Mystery of Carbon
An introduction to carbon materials
Manijeh Razeghi

Chapter 6

Carbon-based polymers

6.1 Introduction

In this chapter, we review the electronic properties of conjugated polymers, in particular doped and undoped trans-polyacetylene (PA). We visit the concept of the Peierls solid. This is a semiconductor formed because the atomic structure of the system relaxes to lower the total energy. The Peierls solid brings along a whole plethora of new physics, which has fascinated researchers for many years. New collective electron–phonon excitations accompany the Peierls transition, in particular the so-called 'soliton and polaron' excitations. The sucessful doping of conjugated polymers has led to the development of the field of conductive polymers. Most conjugated polymers only exhibit short-range order, but here in this chapter we also briefly review the class of highly ordered crystalline polymers called the polydiacetylenes (PDAs). These have fascinated physicists for many years, since the 1980s, because they constitute a unique class of ordered crystalline polymers. Researchers were hoping to find for the PDA, applicable electrical and optical properties combined with a high degree of anisotropy. Some success was achieved but PDAs have so far proven to be very difficult to dope to metallic levels while keeping structural order. This, however, does not mean that new techniques developed by chemists cannot sooner or later resolve this question, and we may finally get highly crystalline metallic polydiacetylene.

6.2 Introduction to conjugated polymers

Let us now consider another type of carbon structure which is of great scientific and technical interest: the class of conjugated polymers. The focus here is on those polymers which have or have potentially conductive or photoconductive properties. The field of polymers is in general too vast to be even touched upon in this short review. Polymers, which form plastics, emulsions or fibers, are perhaps the most important of all carbon complexes used in manufacture. You find some form of polymer in almost every product in everyday use. Most conjugated polymers are

doi:10.1088/2053-2563/ab35d1ch6

disordered. A notable exception is the class of the so-called polydiacetylenes (PDAs) [1], which form excellent single crystals; PDAs will be discussed briefly later on for some special features. In this review we will discuss the polymers which are potentially electrically conductive and exhibit new science. Even in this smaller class just defined, there are too many systems to be reviewed, so we start by looking at the polymers depicted in figure 6.1. The new science in conjugated carbon chains or carbon complexes is mainly associated with the electron–ion coupling. Whereas in many solids, electronic structures are analyzed and computed in the first place by neglecting to first approximation the electron–ion or electron–phonon coupling, and this with good results, in conjugated polymers this approach can be misleading and lead to incorrect conclusions. Let us see why, by taking the most famous examples of all, trans-PA.

The importance of the electron lattice coupling in low-dimensional carbon compounds is demonstrated particularly strongly in the famous polymer trans-PA [3], shown in the top of figure 6.1. We shall now see why.

6.3 Band structure and Peierls relaxation

As hinted above, the lowest energy states of organic molecules with carbon rings and chains involve not just electronic states, but the combined electronic and ionic configurations. Electrons and ions cooperate to find the lowest energy states. This cooperation gives rise to a new science in which the dynamics of electrons and ions must be treated together. For example, in general one finds that the addition of one electron or hole to an organic molecule creates a so-called polaron [4] with a more-or-less important ionic distortion, energy shift, and wavefunction distortion. The

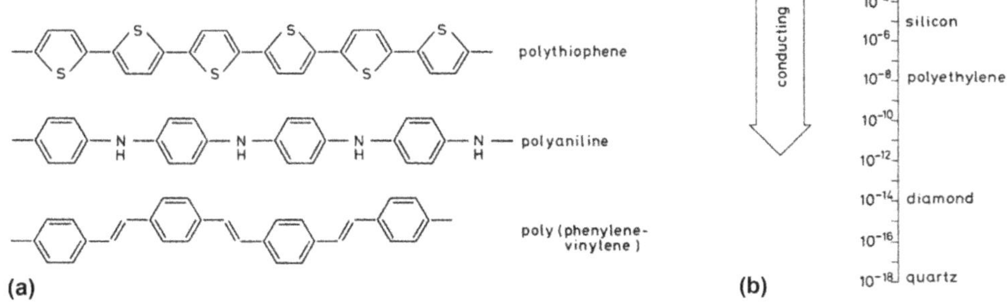

Figure 6.1. Left: some typical conjugated polymers which are of relevance for doping and conduction; starting with trans-PA (top, and right picture): conductivity in S/m chart and comparison. Reprinted with permission from Royal Society of Chemistry, from [2], copyright 1989, permission conveyed through Copyright Clearance Center, Inc.

subject is huge, and the reader is referred to the specialized literature. The new science is particularly striking in low-dimensional linear carbon chains and conjugated polymers. Consider the simplest linear chain polymer, called trans-PA, and shown in figure 6.1. This polymer should, just by looking at it and its chemical structure, have the same C–C bond length along the backbone and then constitute according to the band theory, a metal with a half-filled band. The unpaired pi electrons sticking out of the plane are the ones which form the highest occupied bands. The pi-orbitals have one electron per carbon, which should normally form a conducting band. But this is not what happens: the configuration with the same bond length is not stable. It undergoes the so-called Peierls transition whereby, alternately, one bond forms a pi-double bond and shortens with regards to the other.

Following the work of Su, Schrieffer and Heeger [5], one can take care of the structural bond relaxation by introducing a bond length shift u which gives rise to an ordered alternate short-long bonding arrangement (figure 6.2). Taking into account the exponential distance dependence of the coupling t, $t = t_0 \exp[-\alpha(a - u)]$, on carbon–carbon distance with α denoting the inverse decay length and expanding to first order in u, the ion displacement from apparent equilibrium a, the Hamiltonian with elastic energy $2NKu^2$ included, becomes

$$H_{e+p} = \sum_{i,\sigma} \varepsilon_{i\sigma} c_{i\sigma}^{+} c_{i\sigma} + \sum_{i \neq j,\sigma} t_{ij} c_{i\sigma}^{+} c_{j\sigma} + 1/2 \sum_{n} M\dot{u}_n^2 + 1/2 \sum_{n} K(u_{n+1} - u_n)^2 \tag{6.1}$$

where $\varepsilon_{i\sigma}$ is the diagonal orbital energy at site i and σ the spin index, K the elastic constant, M the ion mass, the $c_{i\sigma}^{+}$ are the creation (+) and annihilation operators for electrons at site i with spin σ. For the shown dimerization and with classical treatment of phonons equation (6.1) becomes

$$H_{e+p} = \sum_{i,\sigma} \varepsilon_{i\sigma} c_{i\sigma}^{+} c_{i\sigma} + \sum_{n,\sigma} [t_0 + (-1)^n 2\alpha u]\left(c_{n+1\sigma}^{+} c_{n\sigma} + c_{n\sigma}^{+} c_{n+1\sigma}\right) + 2NKu^2 \tag{6.2}$$

where M is the ionic mass, the distortion changes a same bond length metallic structure with dispersion $\varepsilon_{k\sigma} = \varepsilon_{0\sigma} + 2t \cos ka$ into a semiconducting one with two atoms per unit cell and a band gap which lowers the electronic energy. The new split bands are given by

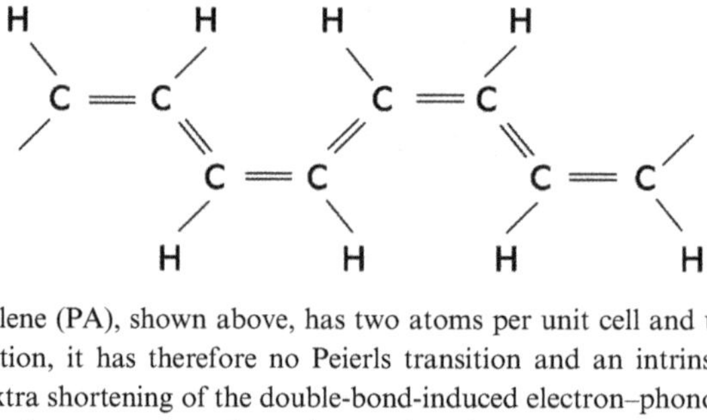

Figure 6.2. Cis-polyacetylene (PA), shown above, has two atoms per unit cell and unequal bond length even without structural relaxation, it has therefore no Peierls transition and an intrinsic band gap. Thus, even though there will be an extra shortening of the double-bond-induced electron–phonon interactions, and single charges will form polarons here, too, the polymer is a semiconductor even without the extra deformation.

$$E_k = \pm\left[\varepsilon_k^2 + \Delta_k^2\right]^{1/2} \tag{6.3}$$

$$\Delta_k = 4\alpha u \sin ka \tag{6.4}$$

where it can be seen that the band gap depends on the strength of the distortion u. As the distortion increases, the valence band gets pushed down in energy. But the distortion also increases the elastic energy by the term shown on the right of equation (6.2), so the amount of distortion is determined self-consistently by minimizing the total energy [5].

6.4 Coulomb interactions in polymers

In the orbital tight-binding language, the most important Coulomb terms are due to the on-site repulsion U between two electrons with opposite spin: the so-called Hubbard Hamiltonian reads

$$H_e = \sum_{i,\sigma} \varepsilon_{i\sigma} c_{i\sigma}^+ c_{i\sigma} + \sum_{i \neq j,\sigma} t_{ij} c_{i\sigma}^+ c_{j\sigma} + \sum_i U_{ii} n_{i\uparrow} n_{i\downarrow}. \tag{6.5}$$

As before, the $c_{i\sigma}^+$ are creation and annihilation ($c_{i\sigma}$) operators for electrons at site i with spin σ. The on-site repulsion U is the energy needed to add an electron to an already occupied orbital, here $n_{i\sigma}$ is the density operator at site i with spin σ (up or down). The magnitude of the bare U for carbon atoms is enormous and ~6 eV, i.e. much larger than any other relevant parameter, including the bandwidth and the electron–phonon interaction. So normally one would have to include this two-body term in the calculations, and, indeed, the simple one-body Hueckel type results and Bloch band structures should be inadequate, or at best very crude approximations. One of the miracles of carbon physics is that this is actually not the case. One can drop the Hubbard term, and indeed the weaker long-ranged electronic coupling terms, as well, and still get an excellent representation of some basic material properties. Justification for dropping the Hubbard term in PA has been given by the work of Kivelson and Heim [6], but their conclusions are strongly based on U not being >1 eV. Surprisingly, it is the much weaker electron–ion coupling which one cannot easily neglect in organic carbon systems, and which at this time is generating much of the new physics. This is a weakness in our understanding and could change once we have more intuitive ways of handling electron–electron correlations. Presently electronic correlations just add to the computational complexity, often producing nightmares without pointing the way to improving design rules. Clearly this has to change, and more effort has to be devoted to tracking correlations as forming a useful part of the material properties. Coulomb repulsion forces particles to correlate their motion in order to avoid each other. Even if it does not explicitly take this into account, parameter renormalization seems to be adequate in many cases. There are exceptions, such as 'Coulomb blockades', and of course ferromagnetism and antiferromagnetism, or metal insulator transitions which rely on strong correlation physics. Recent discoveries with graphene twisted multilayers suggest that order and correlations are closely connected. Slight break in periodicity immediately enhances the role of electronic correlations (see [7]).

6.5 Solitons and polarons in polyacetylene

Doping PA, with n or p impurities, changes the conductivity from hopping conduction to metallic conduction with a concomitant change in the lattice structure. Added carriers do not simply form electrons or holes, but they interact with the already-existing ionic deformation and form new entities called polarons and solitons [1, 3, 4, 8]. This fascinating topic cannot, however, be covered in a short review, and the interested reader is referred to the vast literature on this subject.

The structural degeneracy of the two alternating bonds in the Peierls relaxed structures is shown in figure 6.3. In graphene the dual A–B atom symmetry discussed in chapter 4 has peculiar consequences for the nature of electrons and holes at the Dirac points, inevitably mixing them to acquire a very small effective mass. In the case of PA, when light creates electrons or holes, or when carriers are chemically on the PA chain, they do not just stay as free electrons or holes in the conduction and valence bands, but as a consequence of the structural degeneracy, shown in figure 6.3, the carriers collapse spontaneously into new defect-like states which are in fact elementary excitations of the solitonic type, as shown in figure 6.3, in pairs. A single carrier will partially remove the structural deformation and form a polaronic-type state [3, 4, 8]. This exciting area of carbon science was very topical in the 1980s

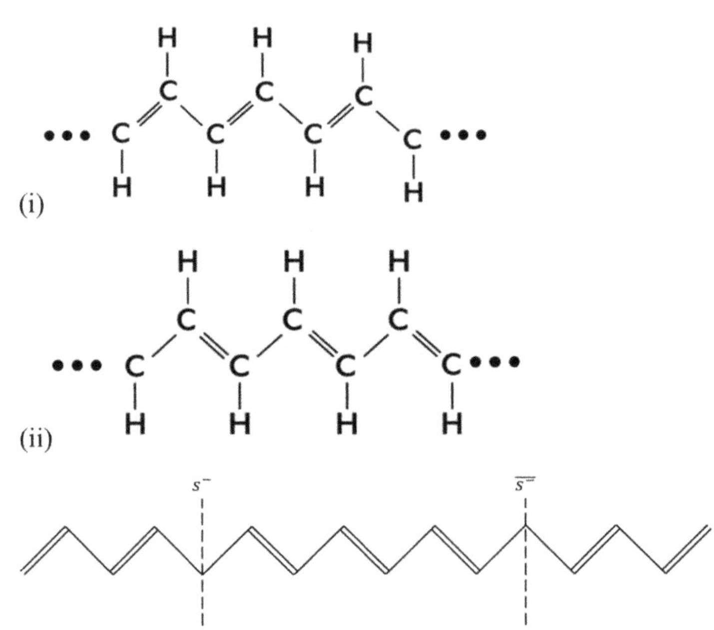

Figure 6.3. Two equivalent bond alternation sequences (i) and (ii) which are energetically degenerate (lower figure) defects described as solitons and antisolitons produced by the phase boundaries between these structures. These defects can be produced spontaneously because they need lower excitation energy compared to exciting a pair of carriers in the conduction and valence band.

and gave rise to a flurry of activity. However, technical exploitation proved difficult at the time because the nanotechnology was not advanced enough, and material stability in air proved problematic. Today we have the tools to handle and manipulate single carriers and even single spins, and we can expect these concepts to come back to some extent. This time around, they will most likely have serious applications in molecule and nanowire switches. They may even form quantum computer hardware. In the majority of non-degenerate conjugated polymers, where this type of structural symmetry does not occur, solitons do not occur, but the polaron concept survives because the electron–phonon interaction is still important. Polarons and bipolarons have been discussed in many reviews.

These midgap states are singly occupied, have spin, and also absorb light. Crosses in figure 6.4 denote data before doping, dotted lines after doping to about 1%. The charge (e, or h) transferred to or from the chains generate spontaneously midgap solitonic defects, as shown in figure 6.4 because these will have lower energy than the quasi-free conduction and valence band states (figure 6.5).

6.6 Conductivity of doped polyacetylene

Polyacetylene can be both n- and p-doped by exposure to vapors of either electron-donating (Li, Na, K) compounds or electron-accepting ones (Br, I, Cl). Doping increases the conductivity of PA in a remarkable way.

The conductivity can be changed by more than 10 orders of magnitude upon doping with a concomitant decrease of the band gap to essentially zero. The low doped temperature dependence of conductivity is strong, and when plotted in a log

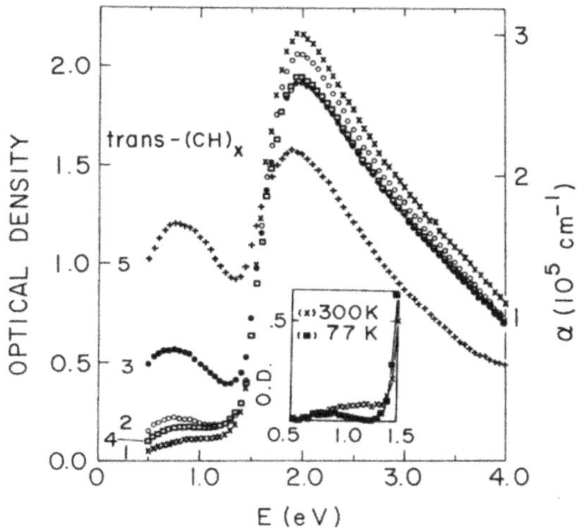

Figure 6.4. Absorption spectra of trans-PA: curve 1 undoped; curve 2 0.01% AsF_5; curve 3 0.1% AsF_5; curve 4 compensated with NH_3; curve 5 0.5% AsF_5. Inset: temperature dependence for undoped trans-PA. The very strong peak near 1.95 eV is due to band-to-band transitions, whereas the hump near 1 eV, which grows with doping and reaches very high values, is taken as strong evidence for the existence of midgap 'soliton defects' S^+ created by doping. Reprinted with permission from [9] © 1980 American Physical Society.

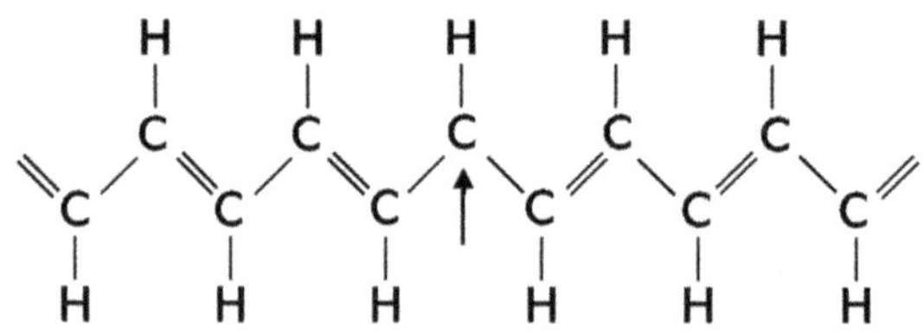

Figure 6.5. The soliton defect in trans-PA. Note that it has spin and can move along the chain with its own characteristic velocity as a nonlinear excitation.

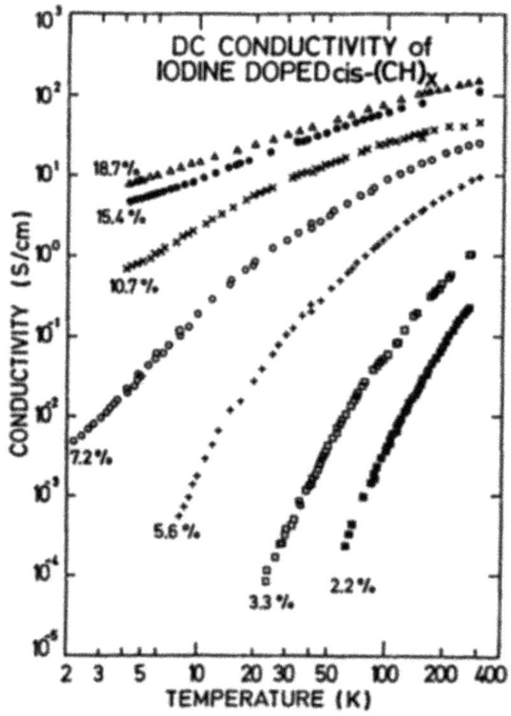

Figure 6.6. DC conductivity of iodine-doped cis-polyacetylene (PA). Reprinted with permission from Royal Society of Chemistry, from [2], copyright 1989, permission conveyed through Copyright Clearance Center, Inc.

scale against $T^{-1/4}$ gives nearly a straight line; this is the signature of Mott cariable range hopping (VRH) conductivity [10]. VRH implies that there is a competition between how far to hop in space and how high the energy barrier is to climb. For mathematical details and derivation of the VRH, see [11].

PA's conductivity can be tuned with several parameters, including the temperature (figure 6.6), the doping level (figure 6.7), or the conjugation length (figure 6.8).

Figures 6.6–6.8 illustrate why PA and related compounds are considered technically important materials. Materials which combine plastic properties, ease of deposition, and have highly tunable conductivities are of value to industry. High thermal conductivity is also an asset but needs special material treatment.

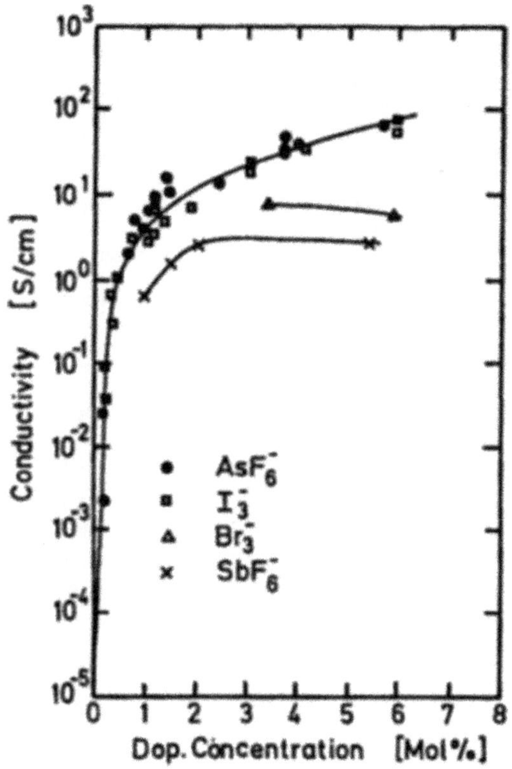

Figure 6.7. Conductivity of trans-polyacetylene (PA) over acceptor doping at room temperature. The high doping regime is truly metallic. Reprinted with permission from Royal Society of Chemistry, from [2], copyright 1989, permission conveyed through Copyright Clearance Center, Inc.

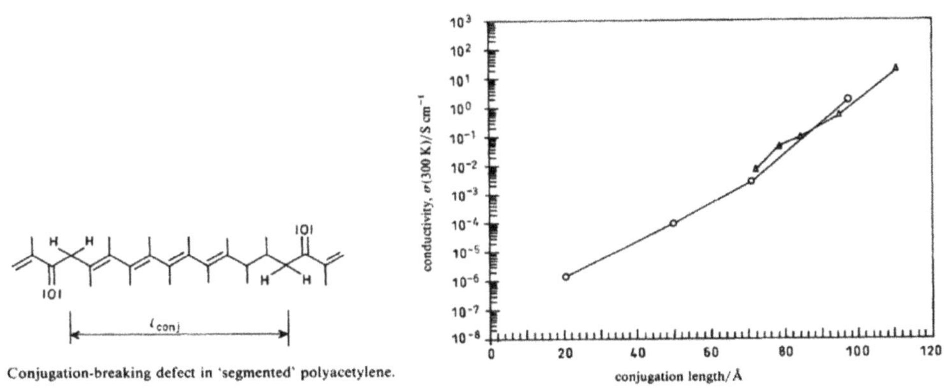

Figure 6.8. Effect of the conjugation length on the conductivity of doped polyacetylene (PA) (right). The conjugation length is changed by inserting the oxygen defect (left). Note the 8 orders of magnitude change. This strong dependence is puzzling but makes such polymer chains interesting for molecular bio-nano-engineering applications, especially sensing and radiation detection. Reprinted with permission from Royal Society of Chemistry, from [2], copyright 1989, permission conveyed through Copyright Clearance Center, Inc.

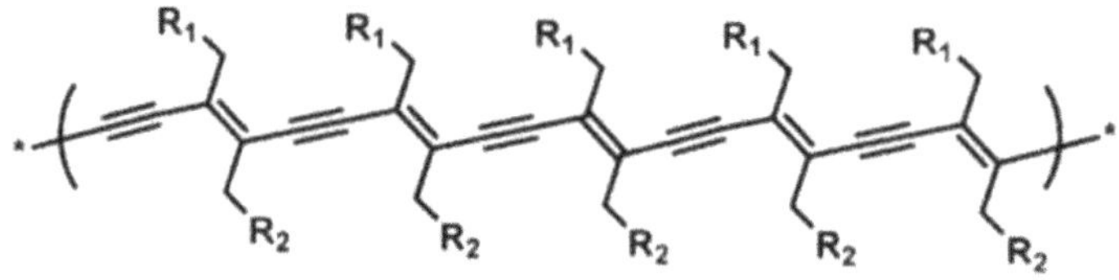

Figure 6.9. The structure of polydiacetylene (PDA). Reprinted with permission from [12]. Copyright 2009, permission conveyed through Copyright Clearance Center, Inc.

6.7 The crystalline highly ordered polymers: polydiacetylene

Almost all conjugated polymers are disordered; crystallinity exists only over finite regions. An exception is constituted by the class of polymers called the polydiacetylene (PDAs). PDAs come in a variety of types which differ from each other by the nature of the sidechains R–R′ attached to the backbone (see figure 6.9).

PDAs have a strong third-order optical susceptibility which makes them interesting for frequency trebling. Frequency doubling is only possible when an external field is applied, which breaks the inversion symmetry [13].

PDAs have been investigated for charge transport along and perpendicular to the chain direction. The anisotropy is enormous: 3–4 orders of magnitude. The pulsed photoconductivity was measured by Donovan and Wilson in a number of papers [25] and more recently analyzed for time decay by Movaghar *et al* [14, 15], where clear proof of one-dimensional motion was given. Both the photoconductive decay and the electric-field-induced second harmonic (EFISH) decay exhibit very long lifetimes, reaching seconds to hours. This suggests that the light does not just produce carrier pairs but leads to serious metastable reconstruction defects which generate long-lifetime hopping charges. Donovan and Wilson have argued that their photocurrent field-dependent data could only be understood if they assumed that the charge dynamic was band polaronic, with a free flight saturated drift velocity of the order of the velocity of sound. A long time of flight was measured by Moses *et al* [16]. They found a mobility of ~5 $cm^2\ V^{-1}\ s^{-1}$, which is the usual equilibrated value expected and found in polymers (figure 6.10).

The reader can verify that PDAs, though not as popular as PAs, exhibit a wealth of new science which is worth pursuing in more detail by going back to the original literature. The new science is a consequence of the unique beautiful crystalline order that one recovers in some of these materials. One can truly speak of molecular wires (see [19]) for truly isolated wires which exhibit strong luminescence – something that the bulk material does not have.

6.8 Bulk heterojunction solar cells based on fullerenes P3HT/PCBM

An exciton is a quasi-particle excitation consisting of a bound electron and hole that mediates the absorption and emission of light, particularly in disordered and

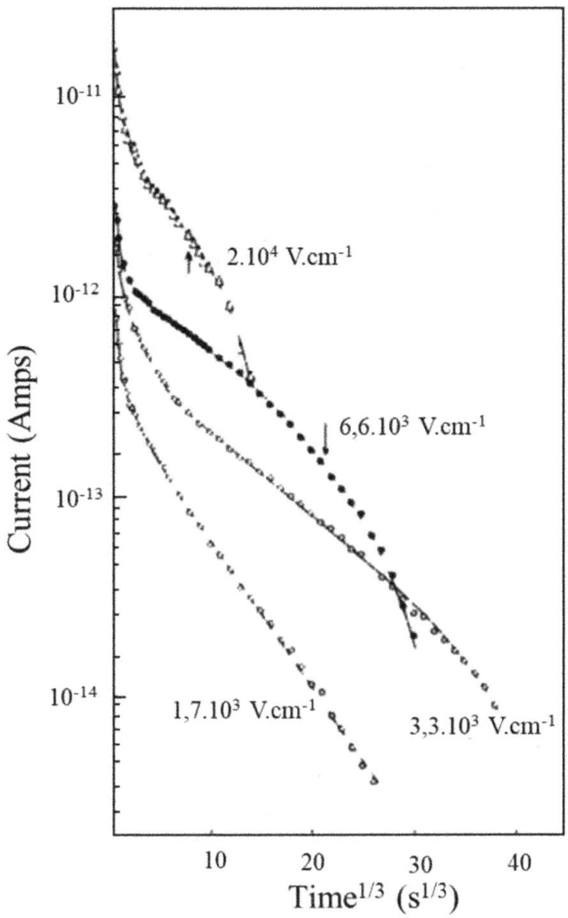

Figure 6.10. Log plot of photocurrent decay versus $t^{1/3}$, the long decay times indicate that the laser pulse generates more than just particle hole pairs. It also produces long-lifetime defects or chain conformation changes, or, indeed, long-time strongly bound polarons which move slowly. Reprinted with permission from [18]. Copyright 1983 by the American Physical Society.

low-dimensional materials. Research on solar cells is now orientated towards excitonic solar cells that do not generate free carriers but rather neutral excitations (figure 6.11).

In this combination of materials, the otherwise difficult generation of free charges out of the exciton precursor is facilitated by the material interfaces (see figure 6.12), which demonstrates why it is favorable by free energy considerations for the exciton to split. The active layer is made of PCBM and P3HT fullerene-based polymers, because P3HT is a good absorber, is ordered and exhibits good hole (and possibly exciton) mobility. PCBM forms also an ideal donor–acceptor interface with P3HT. This polymer therefore absorbs photons; the excitons are diffused to the interface where it dissociates. The carriers diffuse, driven by chemical potential, to electrodes: electrons through the PCBM network to the cathode, holes through the P3HT phase to the anode.

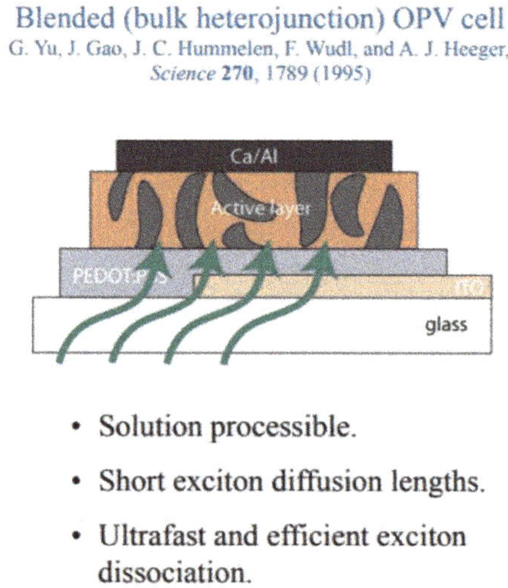

Figure 6.11. Blended (bulk heterojunction) organic photovoltaic cell. Carbon-based solar cells have reached remarkable efficiencies of nearly 15%, not far from amorphous silicon-based solar cells but with better mechanical properties [20].

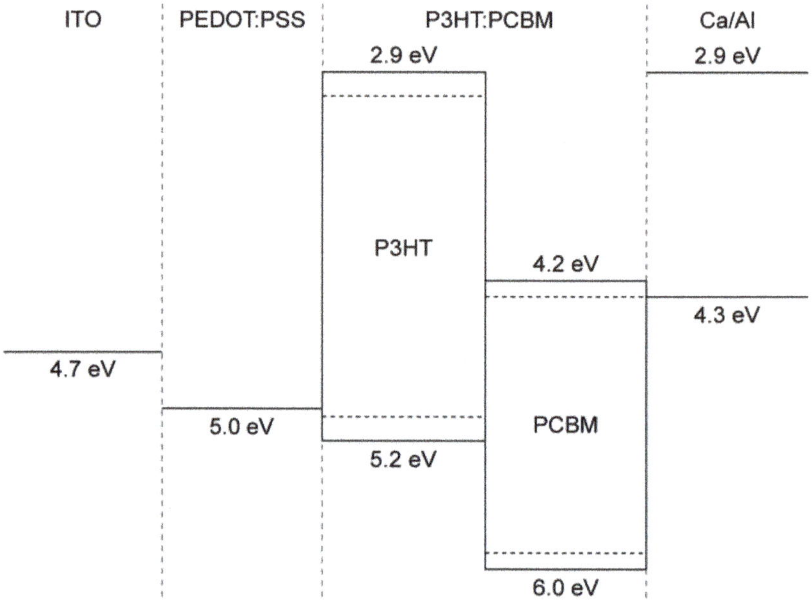

Figure 6.12. Band alignment at the material interfaces of typical P3HT/PCBM solar cell. The exciton created in P3HT breaks up at the interface with PCBM, the electron tunneling and relaxing into the conduction band of PCBM, the hole moves into the poly(3,4-ethylenedioxythiophene) and from there into the contact [20].

The remarkable observation made in these polymer blends is the efficient generation of free charge carriers out of apparently strongly bound excitons generated by light. The otherwise generation of free charges out of the exciton precursors is notoriously difficult in organic materials, making them normally poor photoconductors. In this configuration, charge break up is facilitated by the material interfaces. The light-induced excitons move to the interface during their nano-seconds lifetime and split up into electrons which move into the PCBM (acceptors), and holes which move into P3HT (donor). The dissociation is assisted by the band offset between the conduction band of (for example) PCBM and P3HT (see figure 6.12). How this happens with 100% efficiency on ps time scales, and with apparently no dependence on external bias and temperature, is a subject of hot and ongoing debate [21]. The process does not appear to be the result of many individual consecutive steps, but is due to one quantum mechanical transition to a delocalized eigenstate. If the separation occurs at the interface with energy exchange then this state is called a charge transfer exciton and the energy relaxed by the carriers at the interfaces is lost to heat, at least 0.5 eV; much effort has been devoted recently by chemists and physicists to minimize the interfacial phonon energy loss (figure 6.13).

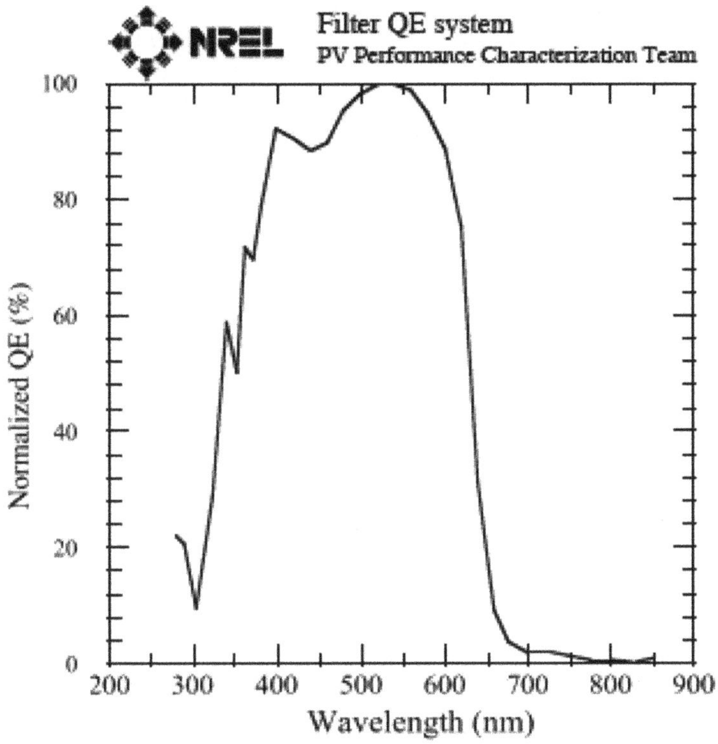

Figure 6.13. Quantum efficiency of a blended (bulk heterojunction) organic photovoltaic cell. Note that the QE reaches 100% in the critical solar radiation region. This is truly remarkable for organic polymers [20].

Bonaccorso *et al* demonstrated in 2015 a dye-sensitized solar cell using graphene as tranparent contact [22]. The light creates excitons on the dyes which then splits up, transferring the electron to the graphene. The electrolyte charge moves to neutralize the positive dyes and picks up electrons from the counter-electrode.

The more modern view about exciton break up, supported by DFT calculations, is that the strongly bound exciton concept is really an isolated molecule effect. In strongly disordered materials such as the above polymer donor-acceptor mixes, and many others, the charge transfer coupling to close neighbors along the edges (~10–80 meV) is a serious effect which allows the pair to break up and delocalize in the network. Again, electronic correlations and order are strongly inter-related.

6.8.1 Tandem organic solar cells

As one can see from the absorption spectra of the best organic blend solar cells, the light harvesting stops in the wavelength region longer than about 900 nm. Researchers have found a way around this by creating tandem solar cell structures which combine two different type of materials, the first one as before [20] the second one chosen to harvest the longer wavelength region [21] (figure 6.14).

6.9 Polymer light-emitting diodes

Typical polymers used in polymer light-emitting diodes PLEDs displays include derivatives of poly(*p*-phenylene vinylene) and polyfluorene. The nature of sidechains inserted onto the polymer backbone can determine the color of the emitted light or influence the stability and solubility of the polymer for performance and affect the ease of processing (figure 6.15).

A typical organic light-emitting diode (OLED) is made of a layer of organic materials placed between two electrodes, the anode and the cathode. These are all deposited on a substrate. The organic molecules can transport electrical charge thanks to the delocalization of the pi electrons, which are the result of conjugation over part or all of the molecule or polymer. As mentioned and reviewed previously, these organic materials have conductivity levels ranging from insulators to (when doped) conductors, and are therefore considered organic semiconductors. Organic semiconductors have highest occupied bands (HOMO levels) and unoccupied bands (LUMO levels), which are analogous to the valence and conduction bands of inorganic semiconductors.

The voltage applied across the OLED is such that the anode is positive with respect to the cathode. Anode materials are picked depending upon the quality of their optical transparency, electrical conductivity, and chemical stability. A current of electrons flows through the device from cathode to anode. The electrons are injected into the LUMO of the organic layer at the cathode and withdrawn from the HOMO at the anode. This depends on the assumption that holes are injected into the HOMO. The electrons recombine with these holes and light is emitted. Electrical forces cause the electrons and the holes to attract and make them move toward each other. When they meet they first recombine to form excitons, which is a bound state of the electron and hole. Since in organic materials holes are in general more mobile

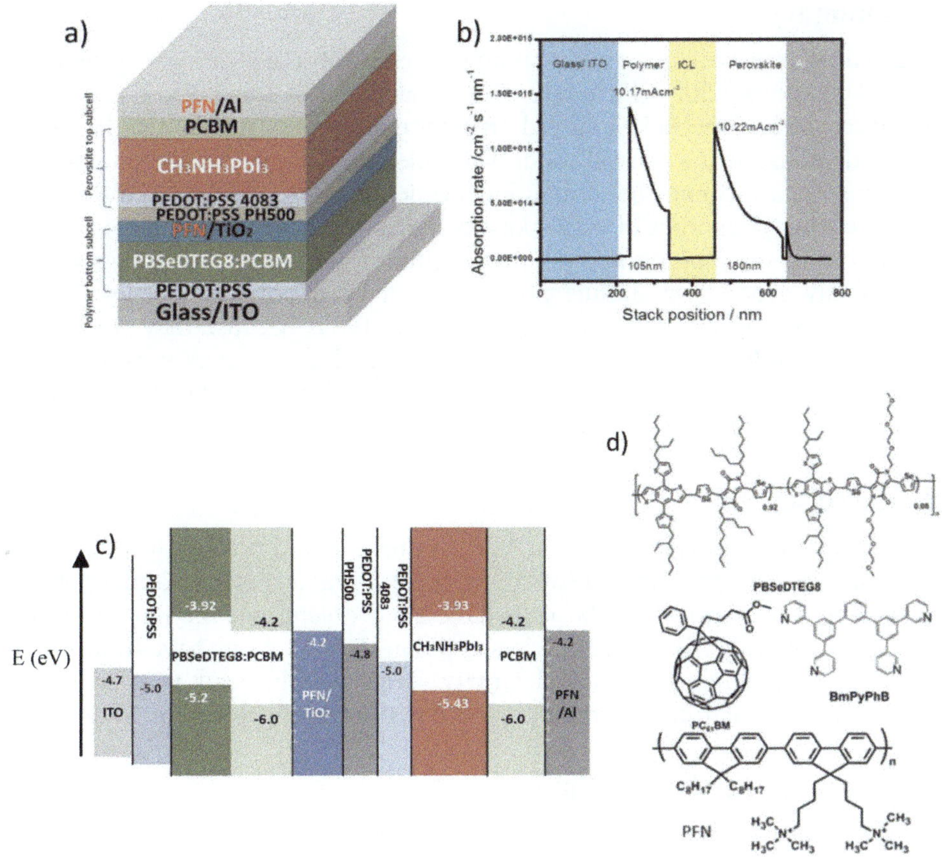

Figure 6.14. (a) Device architecture of the hybrid tandem solar cell composed of perovskite and polymer absorbers; (b) photon absorption rate profile for each subcell inside hybrid tandem solar cell; (c) proposed energy level landscape for the hybrid tandem solar cell. Reprinted with permission from [22]. © The Royal Society of Chemistry 2015.

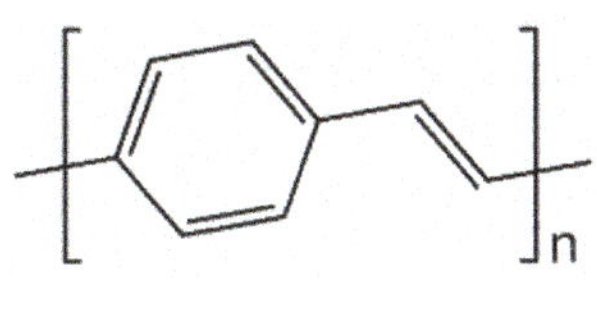

Figure 6.15. Poly(*p*-phenylene vinylene).

than electrons, the exciton formation occurs predominantly closer to the emissive layer. The decay of this excited excitonic state is accompanied by emission of radiation whose frequency is by design in the visible region. The frequency of this radiation depends on the band gap of the material, which is dominated by the difference in energy between the HOMO and LUMO.

6.10 Summary

Conjugated polymers are not only exciting for being 1D-like electronic systems, but constitute materials with which researchers have made some truly remarkable science discoveries. The field enjoyed tremenduous popularity in the 1970s and 1980s and demonstrated a class of solids where electronic structure, lattice vibrations and lattice structure cannot be treated independently. Here we reviewed the doped relaxed electronic properties of trans-polyacetylene, the concept of the Peierls solid, and its accompanying soliton and polaron excitations. We also briefly reviewed the class of highly ordered crystalline polymers called the polydiacetylenes (PDAs). These fascinated physicists for many years in the 1980s because they constitute a unique class of ordered crystalline polymers for which researchers were hoping to find applicable electrical and optical properties, but which proved very difficult to dope. The conduction in PDAs is defect controlled.

6.11 Exercises

(Q1) Explain what is meant by a Peierls transition and how it alters electronic and optical properties. Give an example. What is a soliton and a polaron?

(Q1) Polymer or organic solar cells have the problem that the light excites an exciton rather than free carriers and that these excitons have to be broken up before they conduct electricity. How is this dealt with in current polymer solar cell technology? Why is this not a probem in Si or GaAs?

References and further reading

[1] Pope M 1999 *Electronic Processes in Organic Crystals and Polymers* (Monographs on the Physics and Chemistry of Materials) 2nd edn (Oxford:Oxford University press)

[2] Roth S 1984 Charge transport in conducting polymers *Advances in Solid State Physics* vol 24 (Berlin: Springer)

[3] Etemad S, Heeger A J and Macdiarmid A 1982 Polyacetelyne $(CH)_x$ the prototype conducting polymer *Ann. Rev. Phys. Chem.* **33** 443

[4] Coropceanu V, Cornil J, da Silva Filho D A, Olivier Y, Silbey R and Bredas J-L 2007 Charge transport in organic semiconductors *Chem. Rev.* **107** 926

[5] Su W, Schrieffer J R and Heeger A J 1979 Solitons in polyatcetylene *Phys. Rev. Lett.* **42** 1698

[6] Kivelson S and Heim D 1982 Hubbard versus Peierls *Phys. Rev.* B **26** 4278

[7] Cao Y *et al* 2018 Unconventional superconductivity in magic-angle graphene superlattices *Nature* **556** 43

[8] Bloor D and Movaghar b 1983 Conducting polymers *IEE Proc.* **130** 225

[9] Suzuki N, Ozaki M, Etemad S, Heeger A J and MacDiarmid A G 1980 Solitons in polyacetylene: effects of dilute doping on optical absorption spectra *Phys. Rev. Lett.* **45** 1209

[10] Movaghar B and Schirmacher W 1981 On the theory of hopping conductivity in disordered systems *J. Phys.* C **14** 859

[11] Roth S, Bleier H and Pukacki W 1989 Charge transport in conducting polymers *Faraday Discuss. Chem. Soc.* **88** 223

[12] Yoon B, See S and Kim J-M 2009 Recent conceptual and technological advances in polydiacetylene-based supramolecular chemosensors *Chem. Soc. Rev.* **38** 1958–68

[13] Bloor D, Ando D J, Norman P A, Obhi J S, Kolynski P V and Movaghar B 1987 Electronic and optoelectronic properties of polydiacetylene *Phys. Scr.* **T19** 226–30
[14] Hunt I, Bloor D and Movaghar B 1985 Studies of electric field dependent and temperature dependent charge carrier recombination in one dimension *J. Phys. C Solid State* **18** 3497
[15] Movaghar B, Murray D W, Donovan K and Wilson E G 1984 Anomalous electric field dependence of carrier drift in one dimensional systems. Example organic PDA-TS *J. Phys. C Solid State* **17** 1247
[16] Moses D, Sinclair M and Heeger A J 1987 Carrier photogeneration and mobility in Polydiacetylene: fast transient photoconductivity *Phys. Rev. Lett.* **58** 2710
[17] Cade N and Movaghar B 1983 Polaron states in polydiacetylene *J. Phys.* C **16** 539
[18] Seiferheld U, Baessler H and Movaghar B 1983 Electric field dependent charge carrier trapping in a one dimensional organic solid *Phys. Rev. Lett.* **51** 813
[19] Al Choueiry A *et al* 2010 Twisted polydiacetylene quantum wire: Influence of conformation on excitons in polymeric quasi-one-dimensional systems *Phys. Rev.* B **81** 125208
[20] Rumbles G 2008 Excitonic solar cells: the challenges of efficiency and durability (published by NREL, Golden, CO) https://nrel.gov/docs/fy08osti/43347.pdf
[21] Meng L *et al* 2018 Organic and solution-processed tandem solar cells with 17.3% efficiency *Science* **361** 1094–8
[22] Bonaccorso F *et al* 2015 Graphene photonics and optoelectronics *Science* **347** 6217
[23] Chen C-C *et al* 2015 Perovskite/polymer monolithic hybrid tandem solar cells utilizing a low-temperature, full solution process *Mater. Horiz.* **2** 203–11
[24] Obhi J, Bloor D, Ando D J, Norman P A, Kolinsky P V and Movaghar B 1986 Electric field induced second harmornic generation in the single crystal polydiacetylene PTS-FBS *GEC J. Res.* **4** *Or Physica Scripta Vol T19 226–230, 1987 available on line*
[25] Donovan K J and Wilson E G 1979 Demonstration of high-mobility one-dimensional semiconducting polymers *J. Phys. C: Solid State Phys.* **12** 4857
[26] Bloor D 1985 Polydiacetylene Prototype one dimensional semiconductor *Phil. Trans. R Soc. London* A **314** 51
[27] Bloor D and Chance R R (ed) 1985 *Polydiacetylenes: Synthesis, Structure and Electronic Properties* (Berlin: Springer)
[28] Elman B S, Thakur M, Sandman D, Newkirk M and Kennedy E 1985 Ion implantation studies of polydiacetylene crystals *J. Appl. Phys.* **57** 4996
[29] Yu G, Gao J, Hummelen J C, Wudl F and Heeger A J 1995 Polymer photovoltaic cells: enhanced efficiencies via a network of internal donor-acceptor heterojunctions *Science* **270** 1789
[30] Hummelen J C, Wudl F and Heeger A J 1995 Polymer photovoltaic cells: enhanced efficiencies via a network of internal donor-acceptor heterojunctions *Science* **270** 1789
[31] Zhao W *et al* 2017 Molecular optimization enables over 13% efficiency in organic solar cells *J. Am. Chem. Soc.* **139** 7148
[32] Maurano A *et al* 2010 Recombination dynamics as a key determinant of open circuit voltage in organic bulk heterojunction solar cells. A comparison of four different donor polymers *Adv. Mater.* **22** 4987–92
[33] Heeger A J 2013 25th Anniversary article: bulk heterojunction solar cells : understanding the mechanism of operation *Adv. Mater.* **26** 10–28
[34] Savoie B M *et al* 2014 Unequal partnership : asymmetric roles of polymeric donor and fullerene acceptor in generating free charge *J. Am. Chem. Soc.* **136** 2876

IOP Publishing

The Mystery of Carbon
An introduction to carbon materials
Manijeh Razeghi

Chapter 7

Two-dimensional metal dichalcogenides and their electronic structures

7.1 Introduction

After graphene, researchers tried to find new types of graphene-like two-dimensional layered materials in order to maybe rediscover some of the exciting photonic-like electron band structures. Various groups around the world have found out how to make free-standing barrier layers like hexagonal boron-nitride monolayers, and then finally more recently they discovered how to exfoliate two-dimensional layers of the transition metal dichalcogenides (TMDs). Artificial multilayer fabrication technology is now a very active and popular field of science and technology. We have in this chapter focused only on the truly novel science and applications. Notable discoveries are the ultra-thin film transparent high-ratio field effect transistors (TFTs), which exhibit a high degree of plasticity and are promising for tattoo electronics and many other highly commercial applications. A few potential and unusually exciting science applications are reviewed in this chapter. This includes plastic TFTs and interlayer exciton gas. The strong interlayer exciton binding energy has motivated research groups to try to make interfacial excitonic gases which obey boson statistics and may lead to Bose–Einstein condensates. Such condensates, if realized, would allow a macroscopic number of 'particles' to be accumulated into 'one quantum eigenstate', making it possible to create low-threshold coherence radiation sources on a nanoscale, and maybe even to process quantum information. The coherent condensates could be used to make switches which use interferometry with no loss and no heating.

7.2 Van der Waals and non-van der Waals heterostructures

We have reviewed graphene and its amazing electrical, optical and mechanical properties in chapter 4. After graphene, the next logical step is to investigate bilayer graphene (BLG), and then multilayers. The novelty here is that one is making

doi:10.1088/2053-2563/ab35d1ch7

multiple quantum wells, in analogy with III–V molecular beam-epitaxy-grown semiconductors, but this time the layers are not covalently bonded to each other. They are only van der Waals weakly coupled. The implication and hope is that in such heterostructures, the electronic states in one layer are only weakly perturbed by the other layers, and most of the benefits of graphene can be salvaged, while new ones may be discovered (figure 7.1).

The new energy momentum dispersion of BLG and finite potential is computed with the tight-binding method as

$$E_{\pm}(q) = \left[V^2 + \hbar^2 v_F^2 q^2 + t_p^2/2 \pm \left(4V^2\hbar^2 v_F^2 q^2 + t_p^2 \hbar^2 v_F^2 q^2 + t_p^4/4\right)^{1/2}\right]^{1/2} \tag{7.1}$$

where V is the voltage, q the wavevector, t_p (~0.4 eV) the perpendicular to layer coupling, and v_F the Fermi velocity given by

$$\hbar v_F = 3ta/2 \tag{7.2}$$

where t is the n.n overlap, and **a** the lattice constant ($t \sim 2.5$ eV, a = 0.14 nm) giving a Fermi velocity of 10^8 cm s^{-1} [2].

Figure 7.2 plots the band gap versus charge density doped by a gate field; remember that the effective mass at the Fermi level is also dependent on charge density and thus on the doping level.

We should note that the bilayer structure has not introduced a band gap at the Dirac points. But a slight change in symmetry will produce a gap in particular as noted in chapter 4 [3], a twisting of layers relative to each other changes the band structure creating a superlattice of potentials and superconductivity. But the pairing is not due to phonons, and this example proves for the first time that phonons are

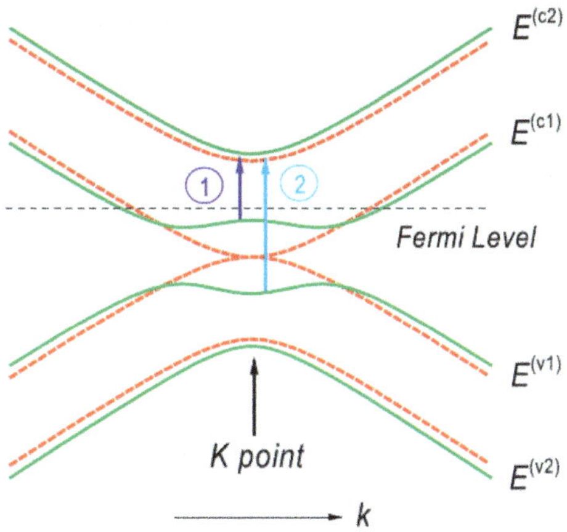

Figure 7.1. Shows the band structure of bilayer graphene (BLG) calculated in the tight-binding approach. You should note that the red lines (no field) exhibit a zero gap behavior at the Dirac (or K,K′) points just like in monolayer graphene (MLG). Transitions 1 and 2 are the strongest optical transitions near the K points for electron doping. Now, if we apply a perpendicular electric field to the layers we have the opening of the band gap (green line). Reprinted with permission from [1]. Copyright 2009 by the American Physical Society.

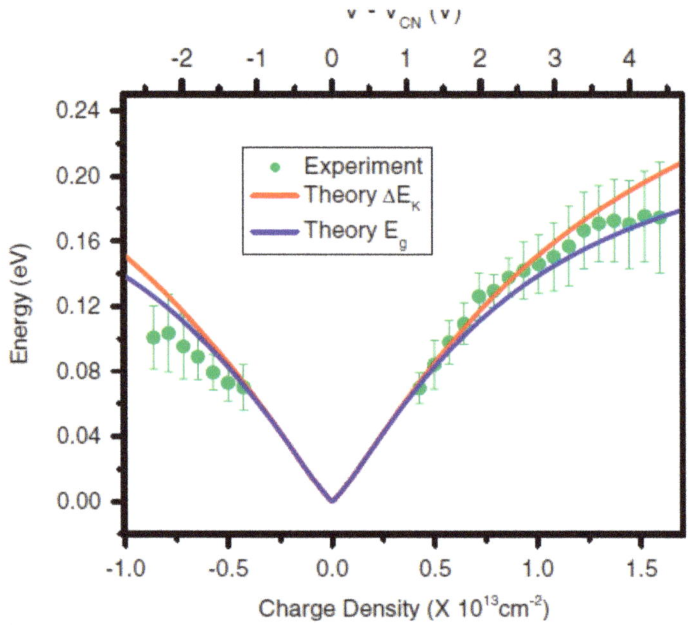

Figure 7.2. Dependence of the energy gap at the K point ΔE_K on the gate voltage and the charge doping density of the graphene bilayer. Note that by grading the charging, one is also grading the band gap, and thus tuning the resistance and photoresponse. Reprinted with permission from [1]. Copyright 2009 by the American Physical Society.

not essential. Multilayer graphene has a rich display of conducting or nonconducting phase transitions [4].

7.3 Transition metal dichalcogenides (TMDs)

Transition metal dichalcogenides (TMDs) are composed of three atomic planes and often two atomic species: a metal and two dichalcogenides. The hexagonal lattice has three-fold symmetry and can permit either mirror plane symmetry or inversion symmetry. This creates non-linear optical phenomena (second harmonic generation, for instance), but most importantly an electronic band structure with direct energy gaps. The metal coordination of layered TMDs can either be trigonal prismatic or octahedral, as shown in figure 7.3.

The electronic structure of TMDs strongly depends on the coordination environment of the transition metal and its d-electron count. The diverse electronic properties shown in table 7.1 arise from the progressive filling of the non-bonding bands from group 4 to group 10 species. When the orbitals are partially filled, TMDs exhibit a metallic behavior, while when the orbitals are fully occupied this leads to a semiconductor behavior (figure 7.4). The effect of the chalcogen atom is minor compared to the metal one, but it is still remarkable: the greater the atomic number, the larger the d band and the smaller the band gap.

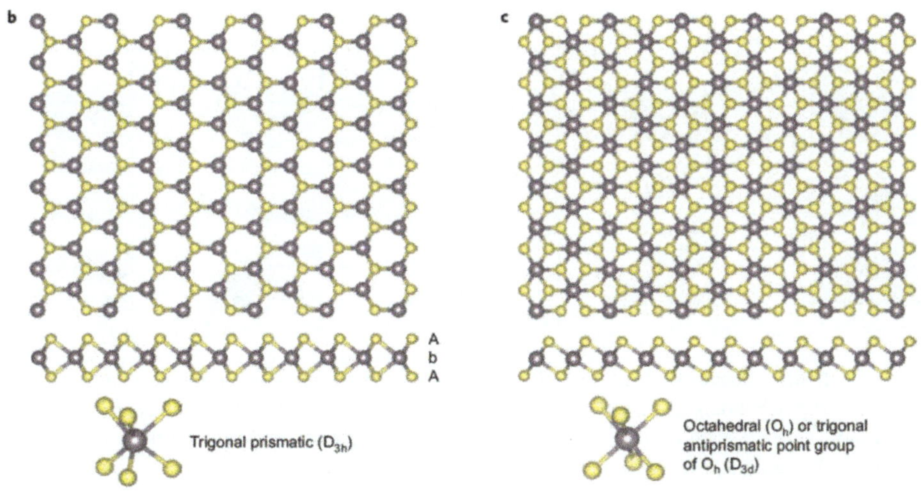

Figure 7.3. Structure of monolayered transition metal dichalcogenide (TMD). The labels AbA and AbC represent the stacking sequence where the upper- and lower-case letters represent chalcogen and metal elements, respectively. Reprinted from [5] by permission from Springer Nature, copyright 2013.

Table 7.1. Electronic character of different layered TMDs[25]. Reprinted from [5] by permission from Springer Nature, copyright 2013.

Grouup	M	X	Properties
4	Ti, Hf, Zr	S, Se, Te	Semicounducting (E_g = 0.2~2 eV). Diamagnetic.
5	V, Nb, Ta	S, Se, Te	Narrow band metals (ρ ~10^{-4} Ω cm) or semimetals. Superconducting. Charge density wave (CDW). Paramagnetic, antiferromagnetic, or diamagnetic.
6	Mo, W	S, Se, Te	Sulfides and selenides are semiconducting (E_g ~ 1 eV). Tellurides are semimetallic (ρ ~ 10^{-2} Ω cm). Diamagnetic.
7	Tc, Re	S, Se, Te	Small-gap semiconductors. Diamagnetic.
10	Pd, Pt	S, Se, Te	Sulfides and selenides are semiconducting (E_g = 0.4 ev) and diamagnetic. Tellurides are metallic and paramagnetic. $PdTe_2$ is superconducting.

ρ, in-plane electrical resistivity.

TMD monolayers are atomically thin materials that exhibit semiconductor properties (table 7.1). They are of the type MX_2, with M being a transition metal atom and X a chalcogen one.

7.3.1 Transition metal dichalcogenide synthesis

TMD sheets have so far been made in several ways, as described in the work of Chhowalla *et al* [5].

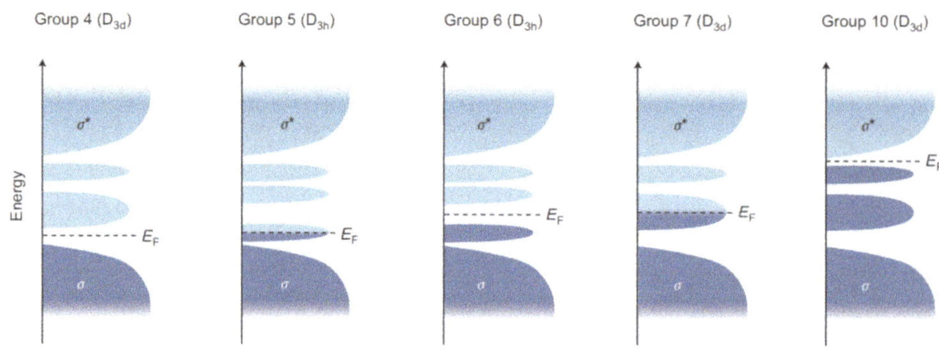

Figure 7.4. Qualitative schematic illustration showing the progressive filling of d-orbitals that are located within the band gap of bonding and antibonding in groups 4, 5, 6, 7 and 10 TMDs. The D_{3h} and D_{3d} refer to the point groups associated with the trigonal prismatic and the octahedral coordination of the transition metal oxides. Reprinted from [5] by permission from Springer Nature, copyright 2013.

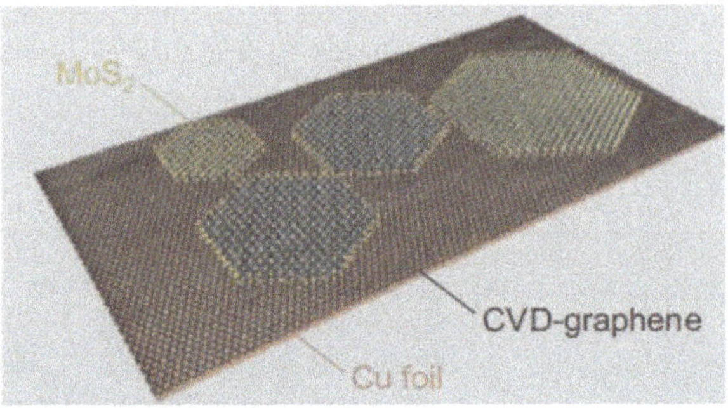

Figure 7.5. Illustrates the quality achievable for heterostructures MoS_2 on graphene grown by chemical vapor deposition (CVD). Reprinted from [5] by permission from Springer Nature, copyright 2013.

(1) Mechanical exfoliation (using scotch-tape), yielding the highest quality monolayered samples.
(2) Liquid exfoliation with selected surfactant with right surface energy penetrating the layers and dissolving them, having a yield of almost 100% and allowing large-scale production of high-quality single layer TMD.
(3) Chemical vapor deposition (CVD) on copper, producing large-area ultra-thin TMD layers; this technique is, however, still not mastered and the uniformity of monolayers is currently a focus of active research (figure 7.5).

7.3.2 Absorption and reflection of transition metal dichalcogenides

Graphene and monolayer TMDs are promising materials for next-generation ultra-thin optoelectronic devices. Although visually transparent, graphene is an excellent sunlight absorber, achieving 2.3% visible light absorbance in just 3.3 Å thickness. TMD monolayers also hold potential as sunlight absorbers and may enable

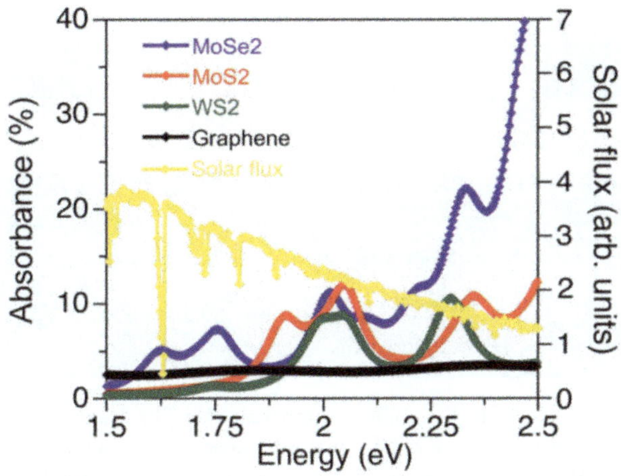

Figure 7.6. Absorbance of TMD monolayers. Comparison of three TMD monolayers and graphene overlapped to the incident solar flux. Reprinted with permission from [6]. Copyright 2013 American Chemical Society.

ultra-thin photovoltaic (PV) devices due to their semiconducting character. Bernardi *et al* [6] demonstrated that the three TMD monolayers MoS_2, $MoSe_2$, and WS_2 can absorb up to 5%–10% incident sunlight in a thickness of less than 1 nm, thus achieving 1 order of magnitude higher sunlight absorption than GaAs and Si (figure 7.6). These calculations unveil the potential of graphene and TMD monolayers for solar energy absorption and conversion at the nanometer scale.

Note by how much the TMD absorption exceeds that of graphene. One monolayer absorbs roughly 10% of the incident light.

7.4 Flexible transistors

We have already demonstrated in chapter 4 the potential graphene has in modern technologies, especially in transistors. However, the absence of a sizable band gap prevents its use in logic circuits and has paved the way for exploration of semiconducting TMDs, which have a huge potential for optical, mechanical, thermal and chemical applications, and of course for electronics. Hexagonal boron-nitride (h-BN) is a large band gap insulator and can be used as a substrate with better interface qualities. Researchers have thus combined the properties of graphene, TMDs and h-BN to shape the future of nanoelectronics. Thin film transistors (TFTs) are a promising opportunity to exploit these two-dimensional materials, even constituting the basics of light-emitting diodes or liquid crystal displays, while finding applications in many other fields. TFTs require high carrier mobility, high on/off current ratio, low contact resistance, presence of both electron and hole conduction, high optical transparency, temperature stability and, of course, mechanical flexibility. Two-dimensional materials are thus natural candidates for this purpose.

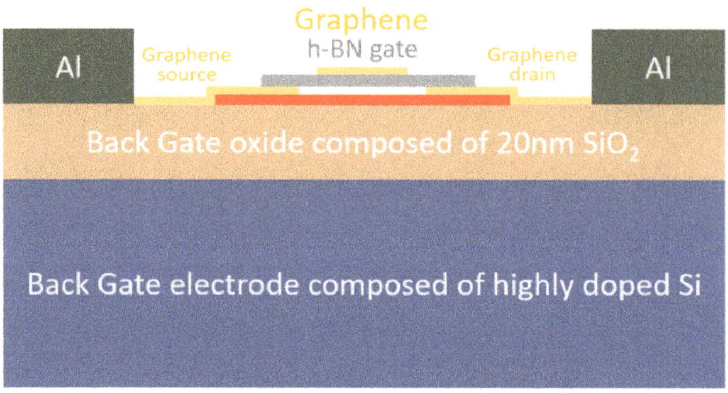

Figure 7.7. Ten-atomic-layer thick, high-mobility, transparent thin film transistor (TFTs) with ambipolar device characteristic. Monolayer graphene is the metal electrodes, 3–4 atomic layers of h-BN was the gate dielectric and a bilayer of WSe_2 was the semiconducting channel for the TFT. The field effect carrier mobility was extracted to be 45 $cm^2\ V^{-1}\ s^{-1}$. Adapted with permission from [7]. Copyright 2014 American Chemical Society.

Saptarshi Das *et al* [7] have demonstrated all two-dimensional TFTs with monolayer graphene as metal electrodes, 3–4 atomic layers of h-BN as the gate dielectric and bilayers of WSe2 as the semiconducting channel material (figure 7.7). The active device stack was found to be 88% transparent over the entire visible spectrum. On/off ratios of 10^7 were observed in all the two-dimensional TFTs, and the carrier mobility was found to be 100 times higher than in amorphous silicon, traditionally used in TFTs (figure 7.8).

7.5 Applications of transition metal dichalcogenide monolayers to molecular sensing

For various applications, including FETs and photovoltaics, being able to dope TMDs is necessary. Both p- and n-type doping can be achieved electrostatically by application of gate voltage, but it requires extra electrodes (gates) that can increase power dissipation and increase area. Another way of doping is through substitution of the transition metal or the chalcogenide of TMDs with appropriate elements, but it usually creates defects inside the TMD structure. Although physisorption of molecules can also lead to a doping effect, they are unstable in nature. Solution-based functionalization of TMD flakes has also been reported to achieve doping, but the functionalization process is generally complicated and time-consuming. Plasma-based doping leads to the formation of defects and significant reduction in mobility. Sarkar *et al* [8] demonstrated the doping of TMDs by metallic nanoparticles (NPs) that can be used in gas sensing (figure 7.9).

7.6 Interlayer excitons

One of the exciting developments associated with layered structures and their tunability is the prospect of making interlayer excitons. Here the electron would reside in one layer, and the hole in another, separated from the first layer by a very

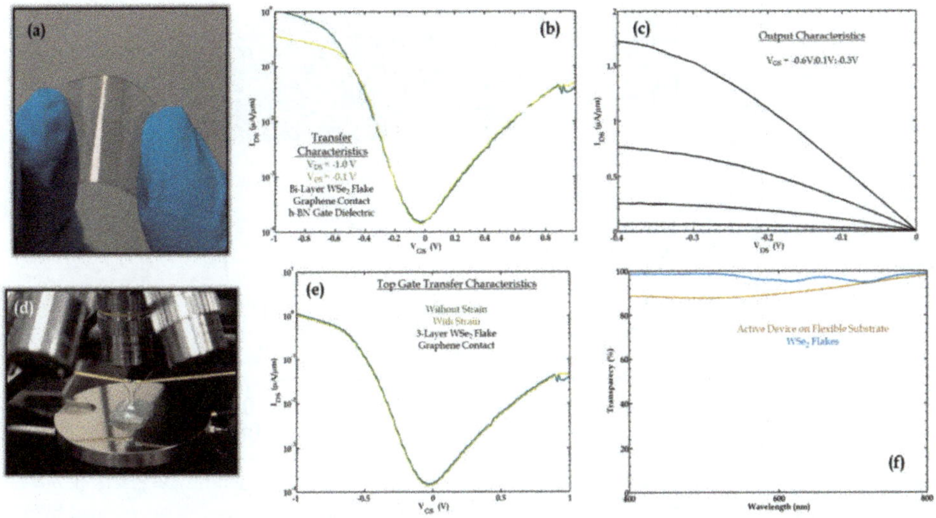

Figure 7.8. All two-dimensional flexible transparent and thinnest thin film transistor. (a) Thinnest field effect transistor on flexible PET substrate, (b) transfer characteristics, and (c) output characteristics of the TFT; (d) experimental setup to measure strain effect, (e) device characteristic under strain, (f) transparency of the TFT and its individual component. Reprinted with permission from [7]. Copyright 2014 American Physical Society.

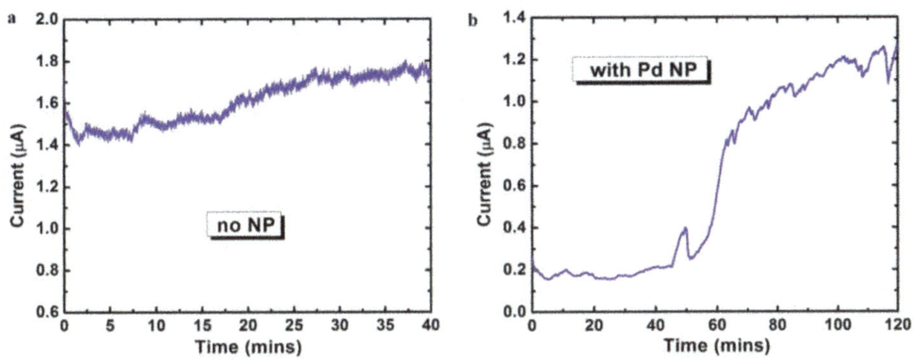

Figure 7.9. (a) Real-time measurement of the current of MoS2 FET without any nanoparticles (NPs). The thickness of MoS2 used was around 8 nm. Negligible change in current was observed when the device was exposed to hydrogen (from time = 20 min onwards). (b) Real-time measurement of current of the same MoS2 FET after incorporation of Pd NPs. Current increases substantially upon exposure to hydrogen (3 ppm from time = 45 min onwards) from 0.2 μA to about 1 μA. Reprinted with permission from [8]. Copyright 2015 American Chemical Society.

thin dielectric barrier. The recent discovery of two-dimensional materials such as TMDs, graphene, and the barrier layer h-BN has generated a new interest in this field. Indeed, with traditional molecular beam epitaxy-grown III–V materials, tuning band gaps and offsets is currently mastered, as is growing layer-by-layer and achieving such interfaces. As a matter of fact, it is in GaAs technology that the first such interlayer exciton gases were made. With type II materials and

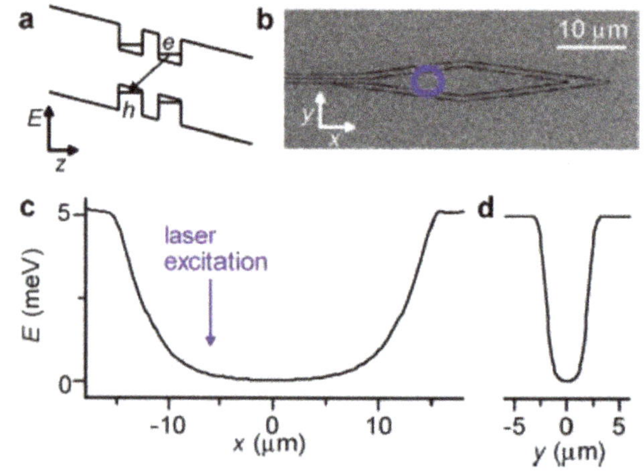

Figure 7.10. (a) Energy band diagram of the 'coupled quantum well' (CQW) structure. e = electron, h = hole. (b) SEM image of electrodes forming the diamond trap; a diamond-shaped electrode is surrounded by a thin wire electrode followed by an outer plane electrode. (c, d) Simulation of exciton energy profile through the trap center along x (c) and y (d). The position of the laser excitation spots is indicated by a circle in (b) and the arrow in (c). Reprinted with permission from [9]. Copyright 2012 American Chemical Society.

superlattices, one could reduce the photo or thermal gap essentially down to zero. Epitaxially grown barrier layers can be as thin as a monolayer (figure 7.10).

Indirect excitons are created in a GaAs/AlGaAs coupled quantum well (CQW) structure. Long lifetimes of the indirect excitons allow them to cool to low temperatures within about 50 mK in an optical dilution refrigerator. This allows the realization of a cold exciton gas with temperatures well below the temperature of quantum degeneracy $T_{dB} = 2\pi\hbar^2 n/m$, with $m = 0.22\ m_0$, $T_{dB} \sim 3$ K, for the density per spin state $n = 10^{10}$ cm^{-2} [10]. By trapping the excitons, evolving Bose–Einstein condensation takes place, runs into the trap, and one can directly observe the coherent light emission with spatial resolution, as shown by High *et al* [9] (figure 7.11).

The coupled semiconducting quantum well interlayer excitons have a very low binding energy (<10 meV) and can only give rise to Bose–Einstein condensation at very low temperatures (<1 K). The potential for Bose–Einstein condensation is what makes the Bose particles so interestig; their otherwise sheer existence is not particularly useful. The small binding energy is a result of the high dielectric permittivity of III–V materials which screens the e–h attraction. This is why the new two-dimensional materials have generated fresh interest. The dielectric constant of h-BN is only about 4, and the exciton binding energy calculated by Fogler *et al* [11] is an astounding 87 meV, i.e. an order of magnitude higher than in GaAs technology. Theoretically speaking, there is now a real possibility of creating a 'high-temperature' Bose–Einstein condensate, which by its very nature (statistics) would have a superfluid behavior. The superfluid matter currents would move with zero viscosity, give rise to quantum interference, and Josephson-like oscillations, a

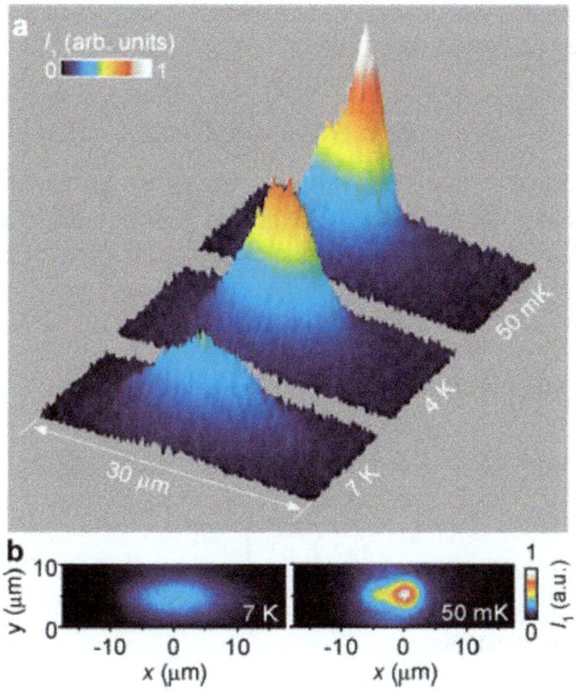

Figure 7.11. (a, b) Emission patterns I_1 for temperature T = 7, 4 and 0.05 K. Light emission from the trap region showing the collective accumulation of matter. The light emitted in the Bose–Einstein phase comes from one condensed wavefunction of many excitons and is coherent. Reprinted with permission from [9]. Copyright 2012 American Chemical Society.

manifestation of the fact that matter waves can accumulate and return in regions of space with almost no cost in energy, in a coherent quantum state. This opens the way for new switching and device applications with far less power consumption and, if realised, would constitute a revolution in technology and computation. Josephson electronics is already a big field but it only works at low temperature.

7.6.1 Bose–Einstein condensation

Bose–Einstein condensation (BEC) is one of the very fascinating phenomena of science. To realize it at practical temperatures is one of the greatest targets and challenges of modern research. BEC occurs in a boson fluid and leads to superfluidity of matter, as in He^4. Current research has focused on achieving the condensation in bosonic systems such as cold atoms and singlet excitons. The idea is that if one can make a condensate at reasonable temperatures, and then control it with electrical and magnetic fields, then one could eventually make computer chips with very low power consumption. Here we briefly review the BEC concept so as to make the remaining discussion understandable.

Consider the Bose–Einstein distribution function $f_b(E)$ which gives us the number of excited bosons in an energy interval:

$$f_b(E) = \frac{1}{\exp(E - \mu)/kT - 1} \tag{7.3}$$

where μ is the chemical potential, E the energy and kT Boltzmann constant times temperature. The chemical potential of a gas of bosons must always be negative otherwise we would obtain a negative particle density at $E = 0$. If we define $g_B(E)$ as the boson density of states then we can write an expression of the total number of particles $n(T)$ as

$$n(T) = \int_0^{\infty} dE \frac{1}{\exp(E - \mu)/kT - 1} g_B(E). \tag{7.4}$$

The density of states for free Bosons has the same form as free electrons:

$$g_b(E) = \frac{p}{4\pi^2} \left\{ \frac{2m}{\hbar^2} \right\}^{3/2} E^{1/2}. \tag{7.5}$$

The degeneracy factor p for $S = 0$ bosons is $p = 1$. Substituting in (7.4), we have for a non-interacting Bose gas

$$n(T) = \frac{1}{4\pi^2} \left\{ \frac{2m}{\hbar^2} \right\}^{3/2} \int_0^{\infty} dE \frac{E^{1/2}}{\exp(E - \mu)/kT - 1}. \tag{7.6}$$

The maximum value of $n(T)$ is obtained when the chemical potential is $\mu = 0$. Now we can actually evaluate the integral analytically to obtain

$$n(T) = 6.4 \times 10^{15} \left\{ \frac{m}{m_0} \right\}^{3/2} T^{3/2}. \tag{7.7}$$

The particle mass has been normalized to the free electron mass m_0.The meaning of this equation is very important: it signifies that ‘if you know the temperature then you also know the maximum number of particles’.

But what happens if we add more particles and keep the temperature constant, or keep the number fixed and lower the temperature? Many physicists would have concluded that there is something seriously wrong with this formalism, but Einstein saw how the program can be ‘repaired’ or, let us say, explained. He noted that maybe the extra added particles can be accommodated into the ground state since there is no limit on how many particles can occupy a given eigenstate in Bose statistics. Writing the total number n_t as

$$n_t = n_0 + n(T) = 6.4 \times 10^{15} \left\{ \frac{m}{m_0} \right\}^{3/2} T^{3/2} + n_0 \tag{7.8}$$

where n_0 is the number in the ground state. One can now define a critical temperature below which the BEC occurs:

$$n_0/n_t = 1 - \left\{6.4 \times 10^{15}\left\{\frac{m}{m_0}\right\}^{3/2} T^{3/2}\right\} \Big/ n_t. \tag{7.9}$$

So we rewrite this as as the following:

$$n_0/n_t = 1 - \left\{6.4 \times 10^{15}\left\{\frac{m}{m_0}\right\}^{3/2} T^{3/2}\right\} \Big/ n_t \tag{7.10}$$

$$(T_c)^{-3/2} = \left\{6.4 \times 10^{15}\left\{\frac{m}{m_0}\right\}^{3/2}\right\} \Big/ n_t \tag{7.11}$$

$$n_0/n_t = 1 - (T/T_c\}^{3/2} \tag{7.12}$$

$$T_c = 2.9 \times 10^{-11}\frac{m_0}{m}(n_t)^{2/3} \tag{7.13}$$

which is the BEC temperature [12].

For a hard core bosonic gas, the scaling is somewhat different, and we have instead of (7.13)

$$k_b T_c = \frac{\hbar^2 c}{a^2 m^*} n_a{}^{2/3}(1 - 0.54 n_a{}^{2/3}) \tag{7.14}$$

where n_a is the number of (hard) bosons per site and c is a constant [13].

The important point about BEC is not that it happens, but that when it happens, completely new properties emerge. The particles all sit in the same wavefunction and are indistinguishable. Thus, the eigenstate acquires macroscopic dimensions and gets macroscopic properties. The condensate is a superfluid: the particles can move without viscosity or without resistance. The eigenstate is protected against scattering and excitation into higher excited states by the BE gap (stimulated down step). In the words of Tony Leggett: 'the bosons all do exactly the same thing!'

But now comes the real question: why does the condensation occur at all? Why is detailed balance apparently not valid below T_c? The reason is not so easy to see and has to do with degeneracy and stimulated emission, like in a laser. When the actual chemical potential is above the ground state, the downward rate into the ground state is enhanced by all the particles in the excited states also trying to get down and increasing the occupation of the stimulating quantum field. Boson–boson scattering emission rates will involve a factor of (1 + nb), which can be very large [14].

So, the flow down a hill into a lower energy state is self-stimulating and has no viscosity or resistance. Considering also plasmon-enhanced Raman scattering, this can be 10^{10} or higher. Here the electromagnetic field is enhanced by the response of the metallic environment and its moving charges.

Making a high-temperature BEC is therefore a highly desirable objective, and this leads us to another theme: excitonic BEC. Singlet excitons have spin 0 and are bosons. They can, in principle, form a BEC, and this has been tried using two-dimensional layers [15].

7.7 Summary

After the discovery of the fabulous treasure chest which is graphene, researchers tried to find new types of graphene-like two-dimensional layered materials. They were finally successful and found free-standing barrier layers like h-BN, and then finally learned how to exfoliate the transition metal dichalcogenides (TMDs) into two-dimensional layers. This has now led to a new revolution in artificial multilayer fabrication technology. The field is in full swing and it is too early to claim that all is known and understood. A few potential and unusually exciting science applications were reviewed in this chapter, including flexible plastic electronics and what could be the possibility of making interfacial excitonic Bose–Einstein condensates (BECs). Such condensates put a macroscopic number of particles into one quantum eigenstate. This makes it in principle possible to process information coherently, i.e. to make switches which use interferometry, working examples being the Josephson junctions of superconductivity.

7.8 Exercises

(Q1) Explain the difference between a free-standing two-dimensional layer and one which is deposited on a substrate, when it comes to electrical and thermal conduction.

(Q2) If a beam of sub-band gap light is passed through a monolayer of MoS2 or bilayer graphene with gate field (i.e. has a band gap), while an electrical current is passed through (an imperfect) the layer, would you expect the light to have an effect on the current? If so, why? If the light energy is just above the band gap?

(Q3) What is the Hall effect? What is so special about the quantum Hall effect? Do some research.

(Q4) Explain the difference between the Bose and Fermi distribution functions.

References and further reading

[1] Mak K F *et al* 2009 Observation of an electric-field-induced band gap in bilayer graphene by infrared spectroscopy *Phys. Rev. Lett.* **102** 256405

[2] Sarma D *et al* 2011 Electron transport in two dimensional graphene *Rev. Mod. Phys.* **83** 407

[3] Cao Yuan *et al* 2018 Correlated Insulator behaviour at half-filling in magic-angle graphene superlattices *Nature* **556** 80

[4] Nam *et al* 2018 A family of finite-temperature electronic phase transitions in graphene multilayers *Science* **362** 324–8

[5] Chhowalla M *et al* 2013 The chemistry of two-dimensional layered transition metal dichalcogenide nanosheets *Nat. Chem.* **5** 263

[6] Bernardi M, Palummo M and Palummo J C 2013 Extraordinary sunlight absorption and one nanometer thick photovoltaics using two-dimensional monolayer materials *Nano Lett.* **13** 3664–70
[7] Das S *et al* 2014 All two-dimensional, flexible, transparent, and thinnest thin film transistor *Nano Lett.* **14** 2861
[8] Sarkar D *et al* 2015 Functionalization of transition metal dichalogenides with metallic nanoparticles : implications for doping and gas sensing *Nano Lett.* **15** 2852
[9] High A A *et al* 2012 Condensation of excitons in a trap *Nano Lett.* **12** 2605
[10] Unuchek D *et al* 2018 Room-temperature electrical control of exciton flux in a van der Waals heterostructure *Nature* **560** 340–344
[11] Fogler M M, Butov L and Novoselov K 2014 High temperature superfluidity with indirect excitons in van der Waals heterostructures *Nat. Commun.* **5** 4555
[12] Peyghambarian N and Koch W 1993 *Introduction to Semiconductor Optics* (Engelwood Cliffs, NJ: Prentice Hall)
[13] Alexandrov A S and Mott N F 1994 Bipolarons *Rep. Prog. Phys.* **57** 1197
[14] Gardiner C W *et al* 1997 Kinetics of Bose–Einstein condensation in a trap *Phys. Rev. Lett.* **79** 1793
[15] Eisenstein J P and Macdonald A 2004 Bose–Einstein condensation of excitons in bilayer electron systems *Nature* **432** 691
[16] Wang Q H *et al* 2012 Electronics and optolectronics of two-dimensional transition metal dichalcogenides *Nat. Nanotechnol.* **6** 699
[17] Qian X *et al* 2014 Quantum spin Hall effect in two dimensional transition metal dichalcogenides *Science* **346** 1344
[18] Dean C R *et al* 2010 Boron nitride substrates for high-quality graphene electronics *Nat. Nanotechnol.* **5** 722–6
[19] Movaghar B 1992 Localisation and the density of states *Phil. Mag. Part* B **65**
[20] Yue *et al* 2013 Adsorption of gas molecules on monolayer MoS_2 and effect of applied electric field *Nanoscale Res. Lett.* **8** 425
[21] Costanzo D *et al* 2016 Gate-induced superconductivity in atomically thin MoS_2 crystals *Nat. Nanotechnol.* **11** 339–44
[22] Rossnagel K and Mingu K 2018 Holstein polaron in a valley-degenerate two-dimensional semiconductor *Nat. Mater.* **17** 654
[23] Alloing M *et al* 2014 Evidence for a Bose–Einstein condensate of excitons *Eur. Phys. Lett.* **107** 10012

IOP Publishing

The Mystery of Carbon
An introduction to carbon materials
Manijeh Razeghi

Chapter 8

Conclusions

The carbon atom features an amazing versatility in its way of bonding with other atoms and itself, enabling the formation of various chemical complexes and allotropes. Studying these astonishing structures has led to promising discoveries, such as the high-temperature superconductivity of potassium-doped carbon fullerenes K_3C_{60}. This unexpected and incredible achievement was highly welcome, as H Kroto was awarded the Nobel Prize. The explanation of this phenomenon still remains partial, and this superconductivity is still not fully understood. The work from Alan Heeger and his Santa Barbara group on trans-polyacetylene (PA) was also an amazing breakthrough in the 1970s. They discovered that this one-dimensional polymer could be both n- and p-doped, just like Su, Schrieffer and Heeger predicted s Peierls semiconductor behavior (chapter 6). In this type of semiconductor, a collective relaxation gives rise to an ordered alternate short–long bonding arrangement, and thus to the formation of an energy gap. Doping PA enables a metallic behavior with a concomitant change in the lattice structure, where the alternating bond length fluctuates back to the same bond length. This interaction between added carriers and lattice structure form new exciting entities called polarons and solitons. Alan Heeger's work was also rewarded by a Nobel Prize in 2000. Possibly the most amazing and exciting discovery was the synthesis of graphene sheets by exfoliating graphite, in the view of material engineering. Consequently, two-dimensional pristine metallic materials could be made and revealed unparalleled structural stability and mechanical worthiness. Geim and Novoselov were again awarded the Nobel Prize in 2010. The suspended form of graphene also exhibited a high mobility (100 000 $cm^2 s^{-1}$), providing nanotechnology a new category of field effect transistor and sensors. Its electronic band structure is also remarkable and unusual: it shows a zero band gap at the gamma point, but also at the Dirac points (figure 4.7), producing linear dispersions, a zero effective mass and a square root magnetic field depending on Landau level splitting. Nanoribbons can be made from graphene and allow a customizable band gap while enabling an electron confinement. Graphene research is only in its infancy and has a great potential for future applications. Interlayer electron–electron coupling and interlayer excitons produce new effects researchers are particularly interested in. In the search for interlayer coupling, the objective is to generate charge polarization in the adjacent layers and then to drag along the induced polarization in the neighboring layers. There is likewise intensive research on Bose–Einstein condensation (BEC) of interlayer excitons using the newly discovered graphene sister materials, the transition metal dichalcogenides (TMDs). In contrast to graphene, the TMDs have a band gap (1–2 eV) but in analogy to graphene, they can also be made in the form of monolayer ultra-thin films and combined with graphene layers to make new superlattice structures. Interlayer excitonic coupling has the advantage of longer exciton recombination lifetimes, which should make it easier to reach the high densities and binding energies needed for BECs (Novoselov). High-temperature superconductivity is also a great challenge. There is some hope that some topology may bring a breakthrough in this area, as with carbon-based high-temperature ferromagnetism. Until now, most of carbon physics has dealt with single-particle mean field theories. While some topologies (such as graphene) suggest a many-body treatment, one-body methods yield surprisingly accurate results and predictions, provided one renormalizes the parameters. Lattice relaxation and phonon interaction are known to be crucial for energy dispersion calculation, but the coupling between electrons is often ignored. These Coulomb interactions, though well presented and quite important, are not well enough understood and apprehended in organic materials. That is why they are traditionally included in the one-body parameters during the renormalization, but there is still a lot to achieve to fully appreciate carbon potential. In potassium-doped fullerenes, superconductivity for instance, there have been several attempts to explain the Hubbard correlation between electrons and to justify the phonon coupling model. This Hubbard correlation would, for example, have an effect around 1.5 eV in Buckyballs. A few researchers have been considering whether it could actually help the formation of bosonic pairs by creating new bands. Understanding the influence of correlations on superconductivity (similarly to magnetism) would be an interesting and promising breakthrough in materials research [1, 2]. The future may reside in unusual topologies, from nanocrystalline quantum dots assemblies to porous forms. Exciting and demanded properties may come from the latest chemical compounds chemists will be able to synthetize, like high-temperature superconductivity [3], lightweight ferromagnetism and diamagnetism, or ferroelectricity. The passivation effect on carbon nanodots has already provided luminescent material, even if all the mechanisms involved are not fully understood yet, while controlling the wavelength seems possible. If the electronic properties of carbon-based compounds are promising, the mechanical robustness of carbon allotropes has found vital applications in civil engineering and transport manufacturing (cars and aircraft mainly). There are still many aspects to investigate and properties to study and combine, for instance in organic solar cells. Developing these applications and research on carbon-based compounds can revolutionize several major industries.

In this review we have focused on carbon as a solid-state material, not as a general organic or biomaterial. These are vast disciplines on their own and are of great interest, especially for biosensing. Organic bioelectronic materials are indeed showing some advantages over traditional microelectronics, such as interconverting ionic and electronic signals, thus efficiently bridging the ionic world of biology with common electronics. Organic materials are also easier to process, and their structural properties allow them to be flexible enough to fit the various shapes organisms have to offer. Conjugated polymeric mixed conductors (both ionic and electronic) improve greatly transduction and amplification of biological signals and are thus widely exploited using poly(3,4-ethylenedioxythiophene) based materials. Graphene has also drawn an increasing interest for use in biochemical sensing applications with its rather unique mechanical and electrical properties. Exfoliation, chemical vapor deposition, electrochemical deposition and solution phase synthesis are the main ways of synthetizing this two-dimensional material, in order to build electrochemical and optical sensors that can detect neurotransmitters, metabolites, proteins, nucleic acids, bacterial cells, and heavy metals. These new devices can open new avenues in healthcare and point-of-care diagnostics. Organic and hybrid perovskite electronic devices hold promise for optical, analytical, and medical applications due to their ease of fabrication, unique simplicity of integration, and potential low cost. Organic light-emitting diodes are one of the best examples of the popularity of these carbon-based devices. Functionalizing graphene sheets can furthermore enhance their mechanical, electrical and chemical properties, and allow graphene sheets to be widely applied in the development of devices for use in fields of industrial, environmental, and biomedical research, and their full potential is only beginning to be developed.

Even as a solid-state review, this document has been kept very brief and simple, and the reader who detects a point of interest should investigate the relevant literature him or herself. The objective here was to show the reader the richness of possibilities that carbon can generate and emphasize some key points and unravel the new science carbon synthesis has uncovered. In engineering technology, automotive, aviation and plastics, carbon, carbon plastics and lightweight fibers are as important as steel and iron. But again, this would be a chapter on its own. The material gathered here is for the benefit of nanotechnology enthusiasts, in particular those involved in searching for applications of quantum phenomena.

References

[1] Hahn J E 2014 Spin triplet s-wave pairing induced by Hund's rule coupling *Phys. Rev.* B **70** 054513

[2] Capone M *et al* 2009 Modeling the unconventional superconducting properties of expanded A3C60 *Rev. Mod. Phys.* **81** 943

[3] Boneschanser M *et al* 2014 Long-range orientation and atomic attachment of nanocrystals in 2D honeycomb superlattices *Science* **344** 1377

Lightning Source UK Ltd.
Milton Keynes UK
UKHW051329190220
358970UK00003B/153

9 780750 311830